Band 19 · November 1983 **11**

isotopenpraxis

AKADEMIE-VERLAG

ISSN 0021—1915 Isotopenpraxis, Berlin 19 (1983) 11, 365—400

EVP 9,— M

ISOTOPENPRAXIS

herausgegeben von J. Mühlenpfordt, Berlin

An unsere Autoren

Die Zeitschrift veröffentlicht Beiträge zu folgenden Themen: Kontrolle und Analyse radioaktiver und stabiler Isotope, Anwendung radioaktiver und stabiler Isotope in allen Gebieten der Forschung und Technik, der Industrie, Landwirtschaft und Medizin, Herstellung radioaktiver und stabiler Isotope sowie von markierten Verbindungen, kernelektronische Geräte für die Isotopenanwendung (Dickenmesser, Dichtemesser u. ä.) Detektoren für isotopentechnische Arbeiten und den Strahlenschutz in Isotopenlaboratorien, Meßtechnik und Strahlenschutz, Dosimetrie in Isotopenlaboratorien, kerntechnische Geräte für die Isotopenanwendung (Beton, Bestrahlungsanlagen u. ä.), Strahlenchemie, Radiochemie, ökonomische und perspektivische Fragen, Fragen der Weiterbildung.

1. Aufnahme von Manuskripten

Zur Veröffentlichung angenommen werden nur druckreife, noch nicht anderweitig publizierte Beiträge in deutscher, englischer oder russischer Sprache zum Themenbereich der Zeitschrift und zwar

- Originalarbeiten
- Kurze Mitteilungen
- Aus der Praxis
- Zusammenfassende Berichte
- Tagungsberichte
- Informationen.

Sind der Beitrag oder Teile daraus bereits veröffentlicht oder ist eine anderweitige Veröffentlichung vorgesehen, so muß dies der Redaktion mit genauer Angabe der Zweitveröffentlichung mitgeteilt werden. Mit der Annahme des Manuskriptes behält sich der Verlag das Recht der Vervielfältigung, Verbreitung und Übersetzung innerhalb der gesetzlichen Schutzfrist vor. Auszugsweiser Nachdruck oder photomechanische u. a. Wiedergabe sind an eine schriftliche Genehmigung des Verlages gebunden. Manuskripte, Zeichnungen und Photos von veröffentlichten Arbeiten werden nur auf den besonderen und bis zum Zeitpunkt der Veröffentlichung mitgeteilten Wunsch des Autors zurückgesandt.

Bei der unverlangten Einreichung von Manuskripten wird empfohlen, eine Befürwortung eines der Herausgeber oder Redaktionsausschußmitglieder beizufügen.

Manuskripte sind einzureichen bei der Redaktion, 1080 Berlin, Leipziger Str. 3–4

2. Form der Manuskripte

Die Autoren werden gebeten:

2 Manuskripte, eines davon in Original-Maschinenschrift, einseitig und mit 2zeiligem Abstand einzureichen; gut leserliche Photokopien sind zulässig; auf dem 1. Blatt (unter Weglassung der Bezeichnung)

den Titel der Arbeit
Vorname-Initialen und Namen sämtlicher Autoren
gegebenenfalls die Institution, in der sie tätig sind
eine kurze Zusammenfassung der Arbeit von höchstens 15 Zeilen
Keywords in alphabetischer Reihenfolge und
als Fußnote die Anschrift der Institution anzugeben;

wenn möglich, auf gesonderten Blättern Namen, Titel der Arbeit und Zusammenfassung in den beiden anderen obengenannten Sprachen 2fach beizufügen;

Literaturhinweise (fortlaufend numeriert), Fußnoten, Tabellen, Abbildungen (in 2–3facher Veröffentlichungsgröße), Abbildungsunterschriften (getrennt von den Abbildungen mit Zeichenerklärungen, Quellenangaben, falls erforderlich, und gegebenenfalls mit einem Vermerk für besondere Wünsche bezüglich der Abbildungsgröße), private und Dienstanschrift der Verfasser mit akademischen Titeln und ausgeschriebenen Vornamen auf besonderen Blättern ans Ende des Manuskriptes zu stellen;

Wünsche für den Kolumnentitel (Seitenüberschrift) bekanntzugeben;
Absätze durch Einzüge von mindestens 3 Anschlägen kenntlich zu machen;
die Maßeinheiten des SI-Systems ("Systeme International") zu verwenden;
Korrekturen im Manuskript unmittelbar an der zu korrigierenden Stelle anzubringen;

Formeln deutlich lesbar und möglichst groß einzutragen;
Farbabbildungen und nachträgliche Änderungen oder Ergänzungen in Abbildungen zu vermeiden;

hervorzuhebende Textstellen zu unterstreichen, und zwar halbfetten Satz mit einer geraden schwarzen Linie, kursiven Satz (Schrägschrift) mit einer blauen Wellenlinie (alle Eigennamen). Sperrungen gestrichelt;
besondere Buchstaben durch farbige Unterstreichung zu kennzeichnen, und zwar Formelzeichen in

Griechisch-Kursiv (α, β, ...) rot,
Kursiv (A, B, ...) blau
halbfetter Kursiv ($\boldsymbol{A}$, $\boldsymbol{B}$, ...) 2 Linien blau (Vektoren),
halbfetter Grotesk (**A**, **B**, ...) 2 Linien violett (Tensoren);

bei Literaturangaben auf Vollständigkeit zu achten. In nachstehender Reihenfolge müssen angegeben sein:

a) bei *Zeitschriften*: Vorname-Initialen und Namen sämtlicher Autoren, Titel der Zeitschrift (nach „Periodica Chimica" oder möglichst wenig gekürzt), Bandnummer (unterstrichen), Jahreszahl (in Klammern), Heftnummer (nur bei Zeitschriften, die jedes Heft mit S. 1 beginnen) sowie die Seitenzahl, auf der die zitierte Arbeit beginnt.

b) Bei *Büchern*: Initialen der Vornamen und Name des Autors (Herausgeber werden durch den Zusatz „ed." gekennzeichnet), voll ausgeschriebener Titel des Buches in der Originalsprache, evtl. Bandnummer, voll ausgeschriebener Name des Verlages, Nummer der Auflage (außer der ersten), Erscheinungsort(e), Erscheinungsjahr (hier nicht in Klammern setzen!) und evtl. Kapitelnummer, Seitenzahl u. ä.

c) bei *Sammelbänden, Tagungsberichten*: Es soll zusätzlich der Titel des zitierten Beitrags angegeben werden. Die Seitenzahl muß unbedingt angegeben werden. Aus dem Zitat muß hervorgehen, ob es sich um einen Tagungssammelband oder um einen noch nicht veröffentlichten Vortrag handelt. Zwischen Tagungsort und Erscheinungsort des entsprechenden Tagungsberichts ist zu unterscheiden.

3. Korrekturen

Der Autor erhält zwei Umbruchabzüge, von denen einer möglichst schnell korrigiert an die Redaktion zurückgereicht werden soll. Auf Wunsch des Autors kann die Korrektur auch von der Redaktion gelesen werden.

Durch nachträgliche Manuskriptänderungen verursachte Korrekturkosten gehen, sofern sie 10% der Satzkosten übersteigen, zu Lasten des Autors.

Der Autor erhält 50 Sonderdrucke kostenlos. Weitere Sonderdrucke können bis zur Rückgabe der Korrekturabzüge auf Kosten des Autors bei der Redaktion bestellt werden.

Bezugsmöglichkeiten

Bestellungen sind zu richten

- in der DDR an den Postzeitungsvertrieb unter Angabe der Kundennummer des Bestellenden oder an den AKADEMIE-VERLAG, DDR-1086 Berlin, Leipziger Straße 3–4, PF-Nr. 1233;
- im sozialistischen Ausland an eine Buchhandlung für fremdsprachige Literatur oder an den zuständigen Postzeitungsvertrieb;
- in der BRD und Berlin(West) an eine Buchhandlung oder an die Auslieferungsstelle KUNST UND WISSEN, Erich Bieber OHG, Wilhelmstraße 4–6, D-7000 Stuttgart 1;
- in den übrigen westeuropäischen Ländern an eine Buchhandlung oder an die Auslieferungsstelle KUNST UND WISSEN, Erich Bieber GmbH, Dufourstraße 51, CH-8008 Zürich;
- im übrigen Ausland an den Internationalen Buch- und Zeitschriftenhandel; den Buchexport, Volkseigener Außenhandelsbetrieb der Deutschen Demokratischen Republik, DDR-7010 Leipzig, Postfach 160; oder an den AKADEMIE-VERLAG, DDR-1086 Berlin, Leipziger Straße 3–4.

Zeitschrift „Isotopenpraxis"

Herausgeber: Im Auftrag des Akademie-Verlages herausgegeben von Prof. Dr. Justus Mühlenpfordt
Verlag: Akademie-Verlag, DDR-1086 Berlin, Leipziger Straße 3–4, PF-Nr. 1233; Fernruf: 2236229; Telex-Nr.: 114420;
Bank: Staatsbank der DDR, Berlin, Kto.-Nr.: 6836-26-20712
Verantwortlich für den Inhalt des Heftes: Prof. Dr. H. Hübner, Leipzig
Anschrift der Redaktion (Fernruf Frau H. Schier: 0351177) der Zeitschrift Isotopenpraxis: Akademie-Verlag, DDR-1086 Berlin, Leipziger Str. 3–4, PF-Nr. 1233
Veröffentlicht unter der Lizenznummer 1527 des Presseamtes beim Vorsitzenden des Ministerrates der Deutschen Demokratischen Republik
Gesamtherstellung: VEB Druckerei „Thomas Müntzer", DDR-5820 Bad Langensalza
Erscheinungsweise: Die Zeitschrift „Isotopenpraxis" erscheint monatlich. Die 12 Hefte eines Jahrganges bilden einen Band. Bezugspreis je Band 180,– M zuzüglich Versandspesen (Preis für die DDR 108,– M). Preis je Heft 15,– (Preis für die DDR 9,– M)
Bestellnummer dieses Heftes: 1079/19/11

AN (EDV) 12831

Isotopenpraxis 19 (1983) 11, S. 365 bis 372

Process Analysis with Nuclear Technique*)

J. Thýn

(Institute for Research, Production and Applications of Radionuclides, Prague)[1])

Experimental determination of response characteristics of individual devices and complete production plants with nuclear technique in different working conditions are discussed. The main principles of the processing of experimental data are described and several examples are presented of measurement from different industrial devices of glass, chemical and pulp industry and of sewage treatment plant

Die experimentelle Ermittlung der Ansprech-Charakteristik von Einzelgeräten und kompletten Produktionsanlagen bei unterschiedlichen Betriebsbedingungen mit Hilfe kerntechnischer Verfahren wird diskutiert. Die wichtigsten Prinzipien der Verarbeitung von experimentellen Daten werden beschrieben; mehrere Beispiele von Messungen aus verschiedenen Industrieanlagen der Glas-, chemischen und Pulpeindustrie sowie einer Abwasserbehandlungsanlage werden vorgestellt.

Keywords

data processing; industrial plants; response functions; tracer techniques

1. Introduction

The stimulus and response methods provides valuable information about linear system behaviour in the form of response characteristic of individual devices or processes i.e. of whole production plants — in time domain (age-distribution functions: residence time distribution — RTD — $E(t)$; $E(\theta)$; the internal age distribution $I(t)$; $I(\theta)$; the intensity function $\lambda(t)$) — in the Laplace transform space (transfer function $E(s)$) — or in the frequency domain (frequency response functions $F(j\omega)$, frequency characteristic, amplitude characteristic $|F(j\omega)|$ = gain factor, and phase characteristic=phase factor). The definitions and the relations are summarized in Tab. 1.

Radioactive nuclides are used with advantage as tracer of the solid, liquid and gas phase or as radiation sources in radiogauging of solid-gas, solid liquid, or the liquid-liquid systems. These are the evident prioritie for workers in service groups, as they cannot be afraid of shortage of their job; on the contrary it may be said that upward tendency in the service work in our country still exists.

The fact, that in radionuclides applications the information mentioned above are practically gained

1) without interferring with the operation (the high sensitivity of detection permits the possibility of adding of small amounts of tracer or in the case of radiogauging no amount is needed);

2) without substantional modifications of devices (the concentration of a component investigated is measured through the walls of devices or pipelines; the detectors are outside of the device) is the dominant priority of this methods.

As it is common in life, the advantages for one are the complications — disadvantages for another one. The engineers-technologists mostly are not willing to get lost these advantages and so the measurement in industry is, from the point of view the experiment realisation, usually complicated even it being carried out in one only device. For example the technologist will not stop the recirculation flow when RTD of device is going to measured and so the experimentator has to use the complicated analysis to obtain information from the responses where impulse response is distorted with the recirculation. It is evidend that experimentator to obtain the same information (RTD f.e.) has to applicate the different techniques for measurement and data proceeding even in the determination of one apparatus response.

The response determination of process, i.e. of complex systems consisting of variously interconnected apparatures may be performed either

1) only experimentally by measurement of the response of the complete system, or

2) by calculation, provided we know for the considered connection of individual devices, their impulse response and the corresponding values of flow volumes, concentrations, etc.

As the response of individual devices is usually not known, it is in both cases necessary to carry out experimental measurements. The direct experimental determination of the response of the complete process is in many cases technically very difficult. In order to be able to record the response of the whole process, doses of the radiotracer would have to be very high, often so high that it would conflict with the safety and health criteria for work with radioactive materials. Individual devices are usually interconnected by means of complex systems of pipes and pumps through which the transported substance may leak into the environment together with radiotracer, and if large activities are involved, surroundings could be radioactively contaminated. Another disadvantage is that we obtain the process response for one arrangement of the individual devices — other variants require a new experiment. The second way presumes a technically simple response measurement of the individual devices. The response of the system is calculated for different operation conditions, i.e. for a cer-

*) Presented at the 2nd Working Meeting "Radioisotope Application and Radiation Processing in Industry", 28th September to 1st October 1982, Leipzig

[1]) Address: 102 27 Praha 10 — Hostivař, Radiova 1. Czechoslovakia

Tab. 1a. Definitions-Relations
Deterministic System

Functions	Definition-Relations	
1	2	
Measured functions Input; Output	$x(t)$; $y(t)$	
Age-distribution x Residence Time Distribution $E(t)$	$E(t) = y(t)/\int_0^\infty y(t)\,dt$; where $x(t-a) = \delta(t-a) = 0$ when $t = a$ $\int_0^\infty E(t)\,dt = 1$	(1a)
$E(\theta)$	$E(\theta) = \bar{t}E(t)$; $\bar{t} = \int_0^\infty tE(t)\,dt$	(1b)
Integral RTD: $F(t)$	$F(t) = \int_0^t E(t)\,dt$	(1c)
Integral age-distribution $I(t)$	$I(t) = 1 - F(t)$	(1d)
Intensity function LAMBDA	LAMBDA $= E(t)/I(t)$	(1e)
Transfer function	$F(s) = L\{E(t)\} = \int_0^\infty E(t) \exp(-st)\,dt = L\{y(t)\}/L\{x(t)\}$	(2)
Frequency response	$F(j\omega) = \int_0^t y(t) \exp(-j\omega t)\,dt / \int_0^t x(t) \exp(-j\omega t)\,dt$; $F(s) \xrightarrow{j\omega = s} F(j\omega) = \lvert F(j\omega)\rvert \exp(-j\theta)$	
Amplitude characteristic (Gain Factor)	$\lvert F(j\omega)\rvert = \{Re^2(F^\circ j\omega)) + Im^2(F(j\omega))\}^{1/2}$	(3b)
Phase characteristic $F(j\omega)$ (Function phase angle)	$\tan\theta = Im(F(j\omega))/Re(F(j\omega))$	(3c)

Tab. 1b. Definitions-Relations
Stochastic system

Functions	Definitions-Relations	
Measured functions Input; Output	$x(t)$; $y(t)$	
Correlation functions		
Auto-correlation f.:	$R_{xx}(\tau) = \lim_{T\to\infty} \frac{1}{T} \int_0^T x(t)\,x(t+\tau)\,dt$; $R_{yy}(\tau) = \lim_{T\to\infty} \frac{1}{T} \int_0^T y(t)\,y(t+\tau)\,dt$	(4a, b)
Cross-correlation f.:	$R_{xy}(\tau) = \lim_{T\to\infty} \frac{1}{T} \int_0^T x(t)\,y(t+\tau)\,dt$	(4c)
Spectral density functions		
One side Power S.D.F.	$G_{xx}(j\omega) = \int_0^\infty R_{xx}(\tau) \exp(-2\pi j\omega\tau)\,d\tau$	(5a, b)
Cross S.D.F	$G_{xy}(j\omega) = \int_0^\infty R_{xy}(\tau) \exp(-2\pi j\omega\tau)\,d\tau$	(5c)
Two sides Power S.D.F	$S_{xx}(j\omega) = \int_{-\infty}^\infty R_{xx}(\tau) \exp(-2\pi j\omega\tau)\,d\tau$; $S_{yy}(j\omega) = \int_{-\infty}^\infty R_{yy}(\tau) \exp(-2\pi j\omega\tau)\,d\tau$	(6a, b)
Cross S.D.F	$S_{xy}(j\omega) = \int_{-\infty}^\infty R_{xy}(\tau) \exp(-2\pi j\omega\tau)\,d\tau$	(6c)

tain throughout and arrangement, from the measured or known response relations of the individual devices. Because in some cases we may presume a certain relation between the response characteristics and the quality of the material produced, we may also estimate the effect of the devices interconnection on the quality of the product.

The determination of response characteristic may be devided in two steps:

1) experimental measurement of the response of individual apparatus;
2) process response calculation.

2. Experimental response determination of Apparatus

The response determination of apparatus need be performed under normal operating conditions of the system. The different experimental conditions required the different kinds of the data processing, which will be mentioned below.

Normalization

For work with radiotracer the impulsed dosage is preferable. This situation is shown in Fig. 1. The RTD values are simply obtained by norming the measured response (see Eq. (1a) in Tab. 1a, for more information see f.e. [1—3], only in case where the input stimulus function may be considered as ideal impulse (i.e. δ-function; where $\int_0^\infty \delta(t)\,dt = 1$). On the Fig. 2 are responses of the ideally mixed vessel, with an impulse of a certain width $W = w/\bar{t}$, that corresponds to the injection duration w. We can see how these are distorted. We shall assume that input stimulus function is an ideal one when the duration of the stimulus is shorter than 5% of the mean residence tims [4—6]. If this condition is not fulfilled, the input stimulus function must be recorded in addition to the response. In this case we obtain the impulse response by analysis of the two recorded curves.

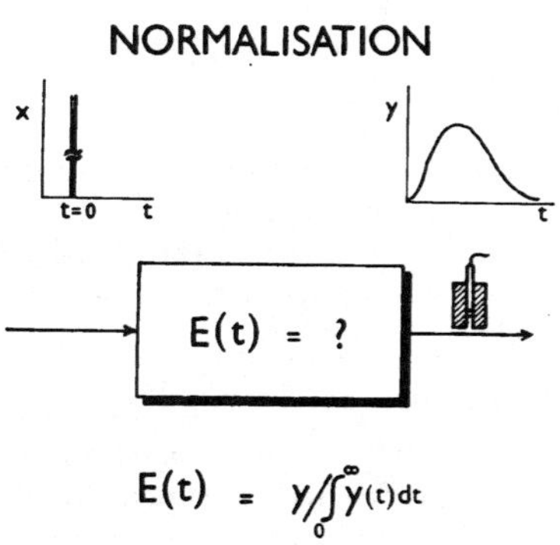

Fig. 1. Impulse response

RTD-Analysis from the response to an arbitrary input function

When the input stimulus function is an arbitrary function-deterministic or stochastic, then impulse response-RTD-$E(t)$ may be calculated from measured input $x(t)$ and output $y(t)$ function. The basic relation which can be obtained from the tracer substance balance [7, 8] is the convolution integral:

$$y(t) = \int_0^t x(u)\,E(t-u)\,du \qquad (7)$$

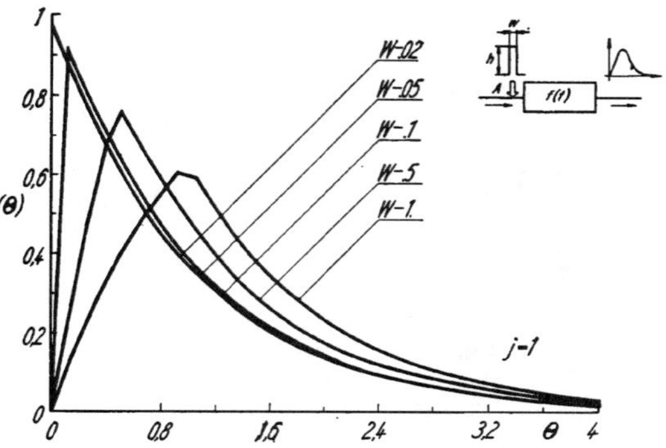

Fig. 2. Rectangular pulse response
$f(\theta) = e^{-\theta}$; $W = w/\bar{t}$

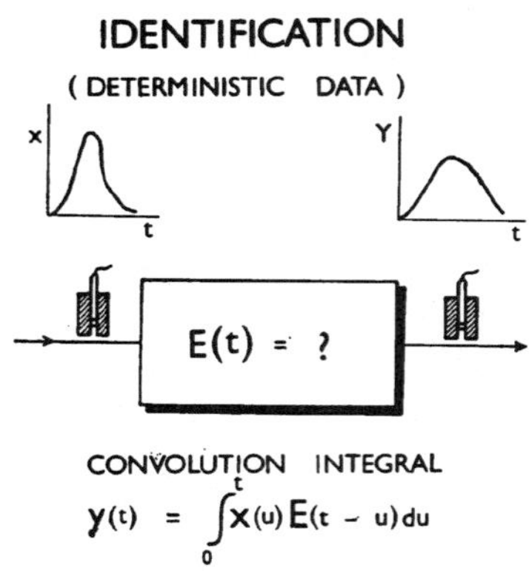

Fig. 3a. RTD-Analysis from response to an arbitrary input function

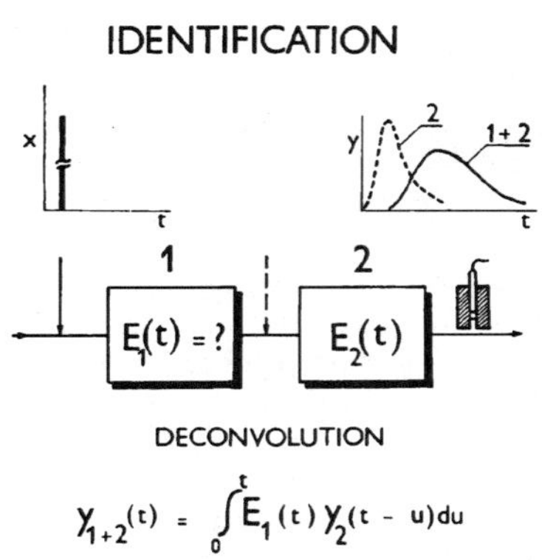

Fig. 3b. Deconvolution

A number of methods were suggested for RTD analysis from integral equation (7) — in the deterministic input signal [9—16] as well as stochastic input signal [17—19].

The response to an deterministic input signal. In the deterministic input signal see Fig. 3a, the identification precision depends on the chosen method, the magnitude of fluctuations of measured data and the properties of the input stimulus function [15]. When the radionuclides are used as a tracer, the measured curves are burdened with fluctuations which must be smoothed before identification, or methods which supress the fluctuations (e.g. the method using the *Laguerre* function (LF) suggested by *Dooge* [12] or the parametric analysis using the Laplace transform, described by *Wen* and *Fan* [3]) have to be utilized in identification. From the comparison of the two latter identification solutions which was done by numerical experiments it follows that the parametric analysis does not yield satis-

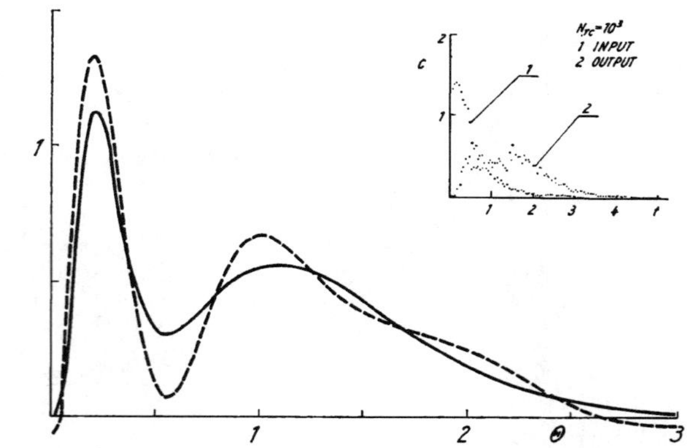

Fig. 4a. Numerical experiment (direct method for identification)

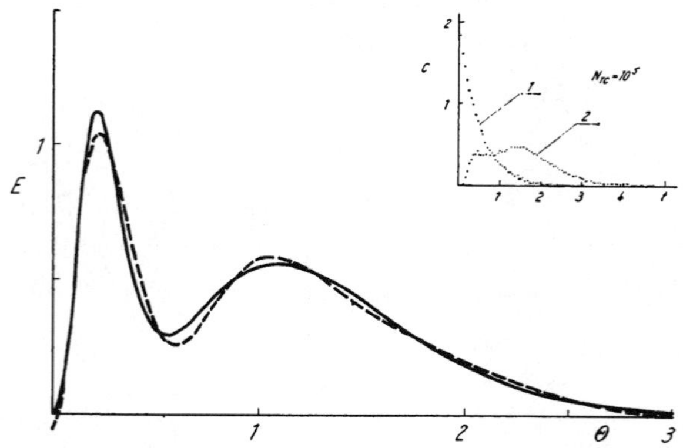

Fig. 4b. Numerical experimental (direct method for identification)

factory results, especially in the analysis of complex RTD i.e. of impulse responses with more extremes [6]. There are two examples of numerical experiments which were performed for the case with small (Fig. 4b) and with greater (Fig. 4a) fluctuation of measured input and output function. From these experiments one may conclude that if the measured curves exhibit fluctuations, the direct method has a better chance of being succesful, provided that the input function has a short duration and its course correspondes to the exponential decay [6].

Generally better results can be obtained when the direct method is used as the first step of the calculation, which is complemented by an optimization method e.g. by the simplex method [16]. The direct method described by *Dogge* [12] and modified for RTD analysis by *Anderssen* and *White* [15] were mostly used in our industrial identification problems.

The numerical algebraic method and this direct method were used when the input function $x(t)$ was the unknown function, where $y(t)$ was measured system response and $E_2(t)$ was the known impulse response (see Fig. 3b). From the comparison of the two methods it follows that the direct method yields better and more precise results than the algebraic one [18].

The response to an stochastic input signal. When the random-stationary-ergodic function is the case of the stimulus input function (see Fig. 5), then from the convolution integral (7) the input-output cross-correlation relation can be found [19]:

$$R_{xy}(\tau) = \int_0^{\infty} E(u)\, R_{xx}(\tau - u)\, du \tag{8}$$

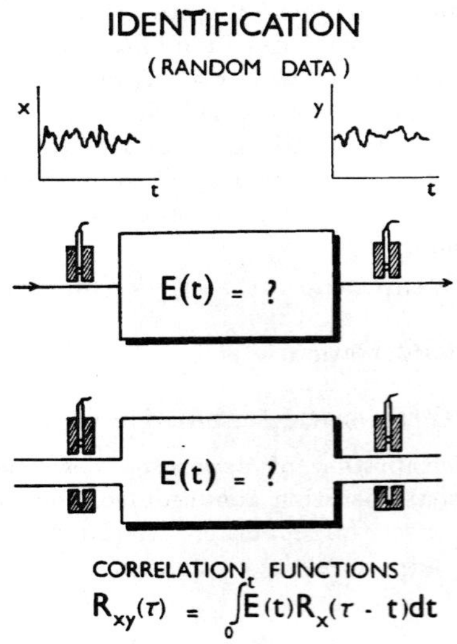

Fig. 5. Random input signal response

where R_{xx} resp. R_{xy} are auto — (see Eq. (4a) in the Tab. 1b) resp. cross — (see Eq. (4c) in the Tab. 1b) correlation function.

There are the infinite limits in the definitions of the correlation functions, and because the realisation of random process $x(t)$ or $y(t)$ are measured in a finite time interval, the estimate of R_{xx} resp. R_{xy} can be only calculated. The estimated autocorrelation function is given by

$$\hat{R}_r = \hat{R}_{xx}(jh) = (n - j)^{-1} \sum_{i=1}^{n-j} x_i x_{i+j} \tag{9a}$$

where n are data values x; $i = 1, 2, \ldots n$,
jh displacement,
j is lag number-time delay, $j = 0, 1, 2 \ldots m$, and
m is maximum value of time delay. (For m, it is recomended $m = 0{\cdot}1n$ [19]).

The estimate of cross-correlation function is calculated by the same formula that is

$$\hat{R}_{xy}(jh) = (n - j)^{-1} \sum_{i=1}^{n-j} x_i y_{i+j} \tag{9b}$$

As the interpretation of the correlation relation (8) in the time domain is not very comfortable, the transformation to frequency domain is used. From the direct Fourier transforms of Eq. (8), so called input-output cross-spectrum relation is obtained:

$$S_{xy}(j\omega) = F(j\omega)\, S_{xx}(j\omega) \tag{10a}$$

where S_{xy} resp S_{xx} are two sides cross resp. power spectral density function which are *Fourier* transform function of cross-resp. auto-correlation functions (see Eq. (6) in the Tab. 1b). $F(j\omega)$ is frequency response function.

The same relation as is Eq. (10a) can be obtained when the one side power and cross spectral density function are used:

$$G_{xy}(j\omega) = F(j\omega)\, G_{xx}(j\omega) \tag{10b}$$

The estimate of G_{xy} resp. G_{xx} can be evaluated by

$$\hat{G}_{xx}(j) = 2h[R_x(O) + 2 \sum_{r=1}^{m-1} D_r \hat{R}_{xx}(jh)] \tag{11}$$

where D_r is so called *Hanning* lag weighting function (for the more details see [19]).

These relations mentioned above were applicated in the segregation evaluation of solid particles discharging flow from the silo, or in the equalization evaluation of the blender [20], and are going to be used in the radiogauging experimental study of the suspension flow in the tubes (for the water transport of the ash from the powerstation).

There is another identification method which seems to be successful for this problem: The maximum likelihood estimation [17]. It enables the determination of the best model for the system as well as the noise model structure. (We have just started to use it and we assume that this method will be successfull also for the identification of the system with deterministic input function.)

RTD analysis from the impulse response of the system working with output stream recycle

When the tracer substance is added so that the input function is an ideal impulse but the outlet stream of the tracer material is partly recycled to the input (see Fig. 6) a special method must be used for RTD analysis. From the tracer balance it follows in this case

$$Qy(t) = AE(t) + rQ \int_0^t x(u)\, E(t-u)\, du \tag{12}$$

where: A amount of tracer added,
r ratio of recycled material ($0 < r < 1$),
Q flow volume

The general solution of integral equation (12) is given in [9, 21] and also approximate solutions have been published [22, 23]. A number of authors [24–31] studied the problem of recycling with respect to RTD.

Good results obtained with the direct method, in the processing of simulated curves with fluctuations [31]

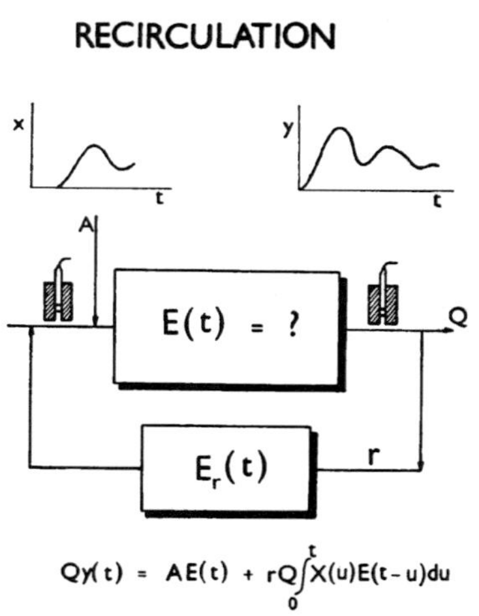

Fig. 6. RTD-Analysis from impulse response of system with flow recirculation

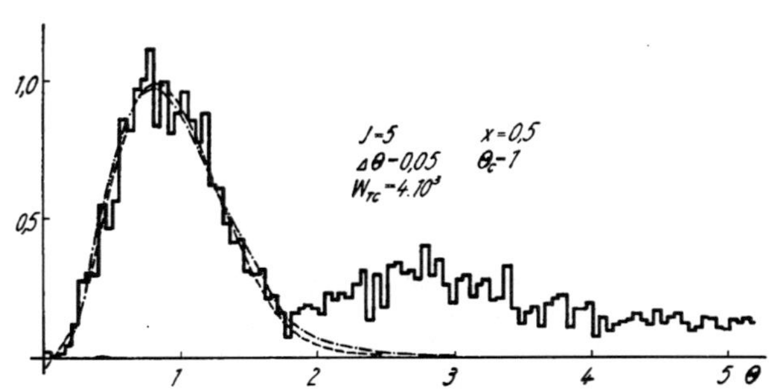

Fig. 7. Numerical experiments
——— recirculation, · · · · · · · theoretical $E(\theta)$, - - - - - $E(\theta)$ calculated by direct method – LF –.

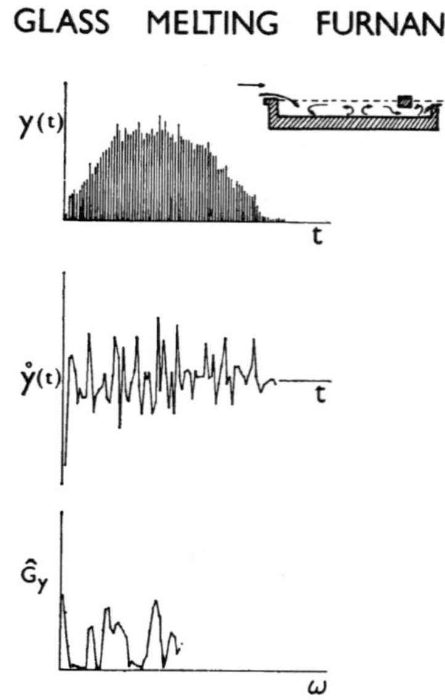

Fig. 8. Floating average filter analysis

indicate usefulness of this method for radiotracer work. The example of the one numerical experiment is shown on Fig. 7. This method was used, therefore, in the treatment of the industrial date (see [32]).

Experimental. Three radiotracer applications of stimulus and response method of one apparatus are describe in short. They are from glass and pulp industry and from the sewage treatment plant.

Our service group carried out more than one hundred experiments in the different glass furnances. The analysis of fluctuations superimposed on the measured impulse response in some cases have been found usefull for the evaluation of glassmelt flow. The impulse response of the glass furnace is demonstrated in Fig. 8. A floating average filter [19] was used to filtering the slow component-residence time distribution. The fast one, this forming fluctuations, was analysed by the method of the random signal. In some cases the dependence of this fluctuation frequence on the radial glassmelt flow in the furnance has been proved; this flow was caused by the reversation of the side heating of the melt. In the evaluation of a glass agregate commonly the balance of power consumption is carried out and the impulse-response measurement is only a part of the extensive experimental measurement.

The applications of the mathematical models of the impulse response – which are built on the base of the physical imagination of the flow system – enable to calculate the impulse response for the different operation conditions. With the use of a mathematical model of impulse response of the activation tank (see Fig. 9a) being aerated with 32 air-inlet tubes, different variances in the distance of purified water input channel for two cases (activation tank with or without sludge regeneration equipment) have been calculated [33]. The results are shown on Fig. 9b. From the several experiments which were carried out in a pulp industry, the evaluation of the vacuum evaporators will be noticed. The six members evaporated station of sulfate liquir were evaluated with radiotracer method. The Fig. 10 shows a lay out of two members and its impulse responses. The responses measured between evaporators at two places of the tube were utilized for liquior flow evaluation. So, the information about evaporated water as

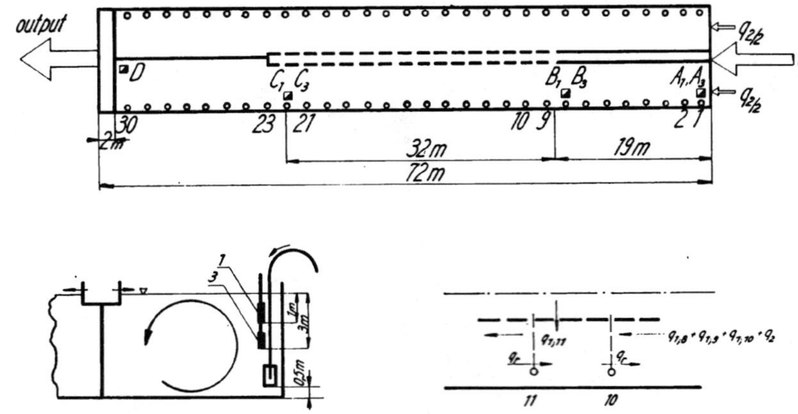

Fig. 9a. Activation tank with 30 air-inlet tubes
1, 3 — scintillation detectors; q_1 — purified water flowrate; q_2 — activation sludge flowrate; q_r — back flowrate

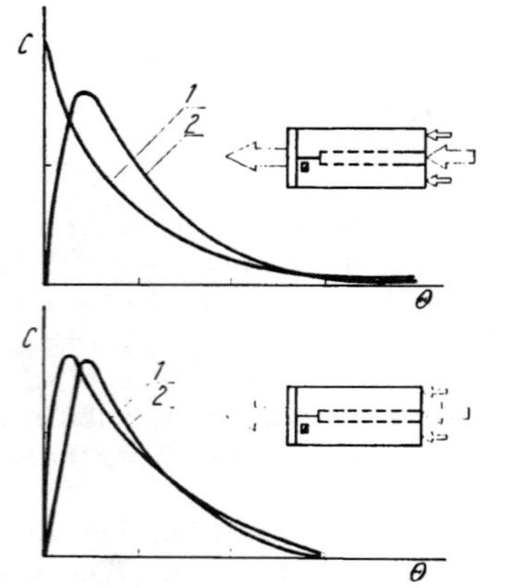

Fig. 9b. Responces in activation thank without or with sludge regeneration equipment
1 — purified water; 2 — sludge

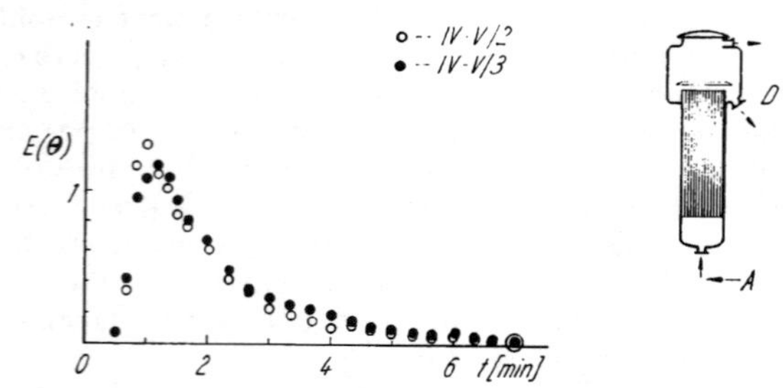

Fig. 10. Sulfate liquir evaporated station
A — radiotracer impulse input; D — detection

well as about the effective volume in the every evaporator could be evaluated.

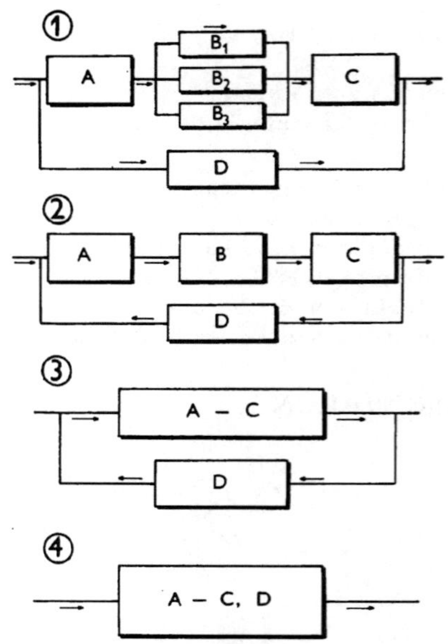

Fig. 11. Gradual simplification method

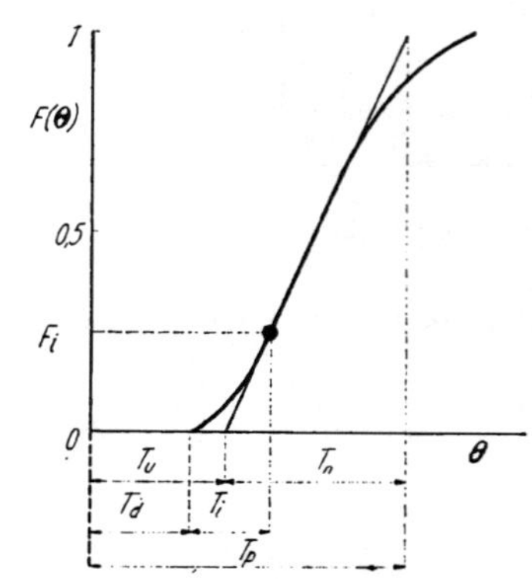

Fig. 12. Transfer function analysis
T_d — time delay, T_n — build up time

3. Process Response Calculation

In the preceding part methods for the determination of the impulse response of single devices were described. In this section we shall deal with the calculation of the impulse response of the production process. The production process is a complex system where the individual devices are connected in series, parallel or in a recycling loop. The response of this system is determined by the actual arrangement of the basic units. Basic units are devices which substantially affect the system response (e.g. large tanks, reactors etc.). For the estimation of the mean residence time of process $\bar{t}_c$, or of the process RTD variance $\sigma_c^2(t)$, the relations shown in Tab. 2a may be used. Relations for precise calculation of the process response characteristics in time or in frequency domain are briefly quoted in Tab. 2b too. This type of processing requires of course a numerical computer.

The calculations may be carried out by the method of gradual simplification described by *Sinclair* and *McNaugh-*

Tab. 2a. Relations for the calculation of process characteristics

Connexion typ	First moment		Second central moment	
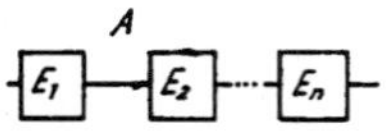	$\bar{t}_c = \sum_i^n t_i$	(13a)	$\sigma_c^2(t) = \sum_i^n \sigma_i^2(t)$	(13b)
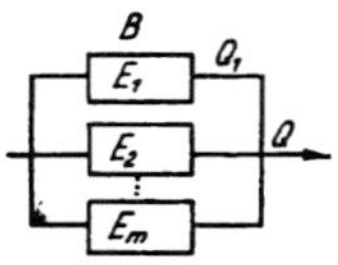	$\bar{t}_c = \sum_i^m Q_i \bar{t}_i / Q$	(14a)	$\sigma_c^2(t) = \int_0^\infty \sum_i^m Q_i (t - \bar{t}_c)^2 \, E_i(t) \, dt / Q$	(14b)
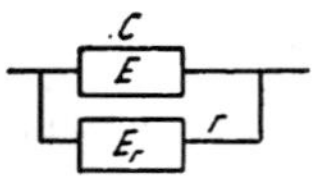	$\bar{t}_c = (\bar{t} + r\bar{t}_r)/(1 - r)$	(15a)	$\sigma_c^2(t) = \sigma^2(t)/(1 - r) + r\sigma_r^2(t)/(1 - r) + r (\bar{t}_c + \bar{t}_r)^2$	(15b)

Tab. 2b. Relations for the calculation of process characteristics

Connection type	Pulse response of the process		Transfer function	
A	$y_c(t) = \int_0^{t_1} \int_0^{t_2} \ldots \int_0^{t_n} x(u_n) \, E_1(t_n - u_n) \, E_2(t_{n-1} - u_{n-1}) \ldots E_N(t_1 - u_1) \, du_n \, du_{n-1} \ldots du_1$	(13c)	$F_c(s) = \prod_i^n F_i(s)$	(13d)
B	$y_c(t) = \sum^m Q_i E_i(t)/Q$	(14c)	$F_c(s) = \sum_t^n Q_i F_i(s)/Q$	(14d)
C	$y_c(t) = (1 - r) \, E(t) + r \int_0^t \int_0^{t_1} E(t - u) \, E(t_1 - u_1) \, y(u_1) \, du_1 \, du$	(15c)	$F_c(s) = (1 - r) \, F(s)/(1 - rF(s) \, F_r(s))$	(15d)

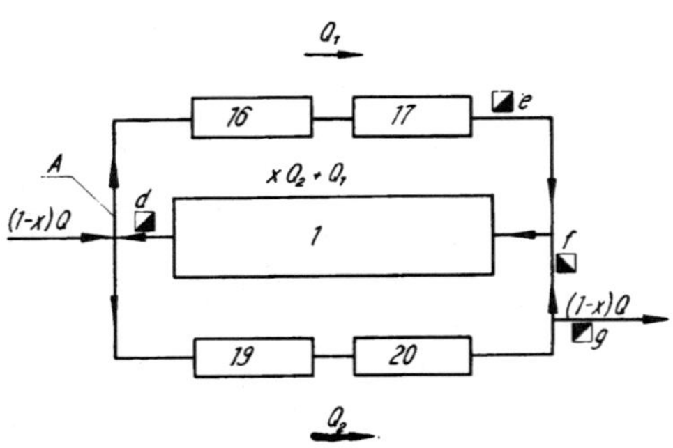

Fig. 13a. Absorption column system
1 — Absorption column with *Rasching* rings; 16, 17, 19, 20 — preheaters; d, e, f, g — scintillation detectors; Q — flowrates; x — recirculation flow

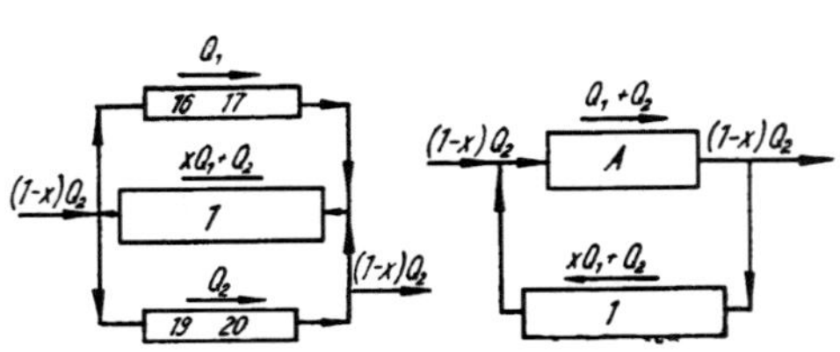

Fig. 13b. Application of gradual simplification method. Absorption column system

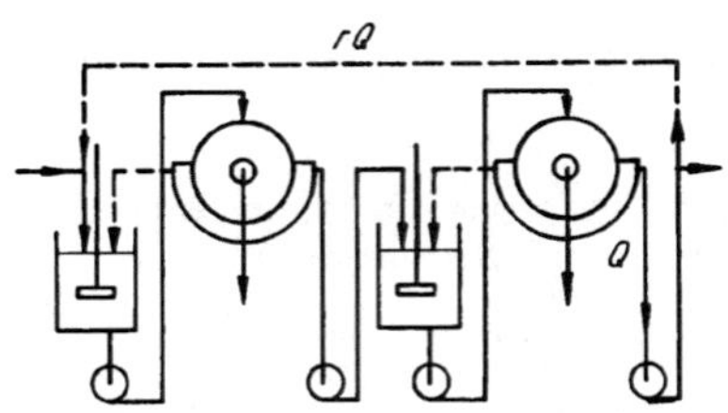

Fig. 14. System of two vacuum filters with floating tanks

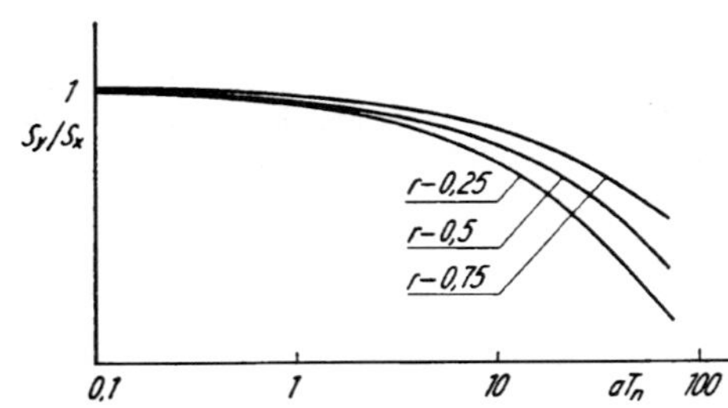

Fig. 14b. The equalisation ability S_y/S_x
r — recirculation flow; a — characteristic value of autocorrelation function of random input signal

ton [35] for example. Fig. 11 shows four steps of such a calculation for a system consisting of four basic units A, B, C and D. The units A, B, C are connected serially, B_1B_2 and B_3 are parallel and the unit D is in the recycling part. If we may presume that $B_{1,2,3}$ are working under identical conditions, then they have equal impulse responses and the calculation scheme is thus simplified: **2**. The response of the system A—B—C for the basic units A, B, C may be calculated from the equation for units connected in series, i.e. from Eq. (13) and the calculus is then simplified to the system A—C with recycle: **3**. Finally, by solving the system with recycle according to equation 15 we obtain the response of the system: **4**.

When the complicated sets of several devices must be analysed to obtain the control system information, then the transfer function is mainly used. The transfer function of one apparatus can be written as

$$F(s) = \exp(-sT_d)/(1 + sT_n) \quad (16)$$

where: T_d time delay,

T_n build up time, time constant

These parameters can be easily evaluated from measured impulse response as it is shown on Fig. 12. The tangent line is put in the inverse point of the integral residence time distribution function $F(\theta)$ (see Eq. (1c) in the Tab. 1a) i.e. in the point of maximum velocity of the concentration change. The process transfer function can be then calculated by Eq. (13d–15d; in the Tab. 2b).

Experimental

Four radiotracer applications of the stimulus and response method have already beem described [32]. Another examples from chemical industry will be mentioned shortly.

In the system of absorption columns the real rate of flow and dispersion of liquid sprinkling the *Raschigs* rings were identified. The example of one column system with four preheaters is shown on Fig. 13a. The evaluation was complicated by the two recirculation flows. The method of gradual simplification was applicated, as it is evident from the Fig. 13b.

The frequence characteristics can be used in the equalizing evaluation of the continuous processes. There is the relation between the auto-correlation function and the standard deviations (or the sample variance s^2), as it follows from the definition of correlation functions (see Eq. (4a, b) in the Tab. 1b). It can be found that for the standard deviation of the output function holds, that [36]

$$s_y = [\int_0^\infty G_{yy}(\omega)\, d\omega]^{1/2} = [\int^\infty |F(j\omega)|^2 \times G_{xx}(\omega)\, d\omega]^{1/2} \quad (17a)$$

When we can suppose that the input concentration $x(t)$ is random function with auto-correlation function given by

$$R_{xx}(\omega) = s_{xy}^2 \exp(-a\tau) \quad (17b)$$

them, after the calculation of G_{xx} by Eq. (5a) the integral (17a) may be evaluated if the gain factor (3b) – that is the real part of the frequence characteristic – is known.

The equalisation ability is given by the ratio of s_y/s_x or by the ratio of s_y^2/s_x^2.

For the system of the two vacuum filters with floating tanks, which is shown on the Fig. 14a, the equalisation ability s_y/s_x as a function of recirculation flow r was calculated. The results are presented on Fig. 14b. The standard deviation ratio is plotted on the vertical axis in the dependence of the characteristic a value of auto-correlation function of random input signal.

Received December 22, 1982
Accepted in revised form March 16, 1983

References

[1] *O. Levenspiel*, Chemical Reaction Engineering. New York: Wiley 1962
[2] *D. M. Himmelblau, K. B. Bischoff*, Process Analysis and Simulation. New York: Wiley 1968
[3] *C. Y. Wen, L. T. Fan*, Models for Flow Systems and Chemical Reactors. New York: Dekker 1975
[4] *J. Thýn*, Radioisotopy **17** (1976) 797
[5] *E. T. J. White*, Imp. Coll. Chem. Eng. Soc. **14** (1962) 72
[6] *J. Thýn, R. Zitný*, Int. J. Appl. Radiat. Isotopes. **26** (1975) 347
[7] *C. W. Sheppard*, Basic Principles of the Tracer Methods. New York: Wiley 1962
[8] *J. Guizerix*, Determination quantitative du degré d'efficacité d'un melaugerur a l'aide de radiotraceurs. Radioisotopes Tracer in Industry and Geophysics Proc. Symp. Prague, 1966 (I.A.E.A., Vienna 1967)
[9] *F. G. Tricomi*, Integral Equations. New York: Interscience 1957
[10] *H. M. Hulburt, S. Katz*, Chem. Eng. Sci. **19** (1964) 555
[11] *R. J. Schwartz, B. Friedland*, Linear System. New York: McGraw-Hill 1965
[12] *J. C. Dooge*, I. J. SIAM Control **2** (1965) 396
[13] *J. H. Moser, C. R. Cupit*, Chem. Eng. Prog. 62/6/60-65 (1966)
[14] *J. E. Gwyn, J.* AIChE, 15/1/126-127 (1969)
[15] *A. S. Anderssen, E. T. White*, Can. J. Chem. Eng. **47** (1969) 288
[16] *I. Holoubek, J. Thýn, V. Varhola*, VII. national conf. Application of radionuclides in technical praxis, Bratislava. 1-3/9/82
[17] *P. Eykhoff*, System Identification. New York: Wiley 1974
[18] *J. Palige, J. Thýn*, Radioisotopy **18** (1977) 273
[19] *J. S. Bendat, A. G. Piersol*, Random data: Analysis and Measurement Procedure. New York: Wiley 1971
[20] *J. Thýn*, Isotopenpraxis **19** (1983) 411
[21] *K. Yosida*, Lectures on Differential and Integral equation. New York: Interscience 1960
[22] *S. F. Mikeladze*, Izv. Akad. Nauk UCCP **2** (1935) 257
[23] *V. T. Krylov*, Trudy mat. Inst. Akad. Nauk UCCP **28** (1949) 33
[24] *van de Vuse*, Chem. Eng. Sci. **21** (1966) 611
[25] *A. Hadad, D. Wolf, W. Resnick*, Can. J. Chem. Eng. **10** (1964) 216
[26] *R. M. Voncken, D. B. Holmes, H. W. Hartog*, Chem. Eng. Sci. **19** (1964) 209
[27] *D. Wolf, W. Resnick*, Ind. Eng. Chem. Fundamentals **2** (1966) 287
[28] *B. Fu, H. Weinstein, B. Bernstein, A. B. Shaffer*, IEC Process Des. Develop. **10** (1971) 501
[29] *T. Sjosted*, Acta polytech. Scand. Ma **17** (1969)
[30] *L. G. Gibilaro*, Chem. Eng. Sci. **26** (1971) 299
[31] *J. Thyn, L. Hansson*, Int. J. Appl. Radiat. Isot. **26** (1975) 748
[32] *J. Thýn*, Int. J. Appl. Radiat. Isot. **32** (1981) 733
[33] *J. Thýn, H. Ulrich, K. Koping, B. Pechlák, L. Blaha*, Radioisotopy **23** (1982) 357
[34] *J. Thýn, B. Pechlák*, Service report UVVVR Prague 1981
[35] *C. G. Sinclair, McNaughton*, Chem. Eng. Sci **20** (1965) 261
[36] *K. Graichen, I.* Chem. E. Symp. series No. 65, Birmingham, 10–13 March 1981

Isotopenpraxis 19 (1983) 11, S. 373 bis 376

Studies on Reduction Process of Cu_2O in an Electric Furnace Using Radiotracers*)

Z. Bazaniak, T. Klimkiewicz

(Institute of Nuclear Chemistry and Technology Warszawa, Poland)[1])

Reduction process of Cu_2O in an electric furnace has been investigated by means of radioisotope technique. Cu-64 in various chemical forms was used as a tracer. Radiometric measurements of samples and complementary chemical quantitative analyses allowed to determine degree of reduction and decoppering degree at all stages of reduction process. Also material balance has been drawn up and distribution of copper among the products determined. On the grounds of the data obtained the possible mechanism of the decoppering process and its kinetic parameters have been discussed.

Der Prozeß der Reduktion von Cu_2O im Elektroofen wurde mit Hilfe radioaktiver Isotope untersucht. Dabei wurde ^{64}Cu in verschiedenen chemischen Formen als Tracer verwendet. Die radiometrischen Messungen der Proben und die ergänzenden quantitativen chemischen Analysen ermöglichten die Ermittlung des Reduktionsgrades und des Entkupferungsgrades in allen Stufen des Reduktionsprozesses. Weiterhin wurde das Materialgleichgewicht ermittelt und die Kupferverteilung über die Produkte festgestellt. Auf der Grundlage der erzielten Daten wurden der mögliche Mechanismus des Entkupferungsprozesses und seine kinetischen Parameter diskutiert.

Keywords

copper; copper 64; industrial plants; metallurgy; radioisotopes; tracer techniques

1. Introduction

Nuclear methods have found wide application for some physical quantities measurements (density, moisture, thickness, etc). The measurements performed by means of radioisotope technique are usualy applied for technological processes control. Sealed radioactive sources are in the common use for analysing the composition of various materials.

On the other hand, application of the free radiation sources for the processes investigation still meets some opposition and objections; caused mainly by fear for the negative influence of the radiation and also by high technical requirements of the radiological safe-guard.

The simplest way of the free radioactive sources application is determining the system dynamics, which requires merely observing the likeness of physical properties of a radiotracer and a labelled material.

However, the works aiming to determine a chemical reactions mechanism or kinetics are limited to a relatively narrow range of processes and methods. Within the attainable range are the reactions involving the elements, which can be transformed by means of neutron activation into radionuclides possesing nuclear properties suitable for measurements.

This work, being an attempt of explaining a mechanism of decoppering process, is an example of applying the chemical tracers.

The technological optimization improving the economic efficiency of the process, consists in the intensification of the reduction process and minimizing irretrievable copper losses in the waste-slag.

*) Presented at the 2nd Working Meeting "Radioisotope Application and Radiation Processing in Industry", 28th September to 1st October 1982, Leipzig

[1]) Address: 03-195 Warszawa, Poland, Dept. XVI, ul. Dorodna 16

2. Experimental

Some of the technological processes being a subject of complex investigations are highly inconvenient for physicochemical measurements due to unfavourable conditions existing in separate units or in the whole installation.

High temperatures and pressures as well as high flow rates of the media make the conventional methods fail completely. In such cases, for investigating the flow dynamics and mechanism or kinetics of chemical reactions radiotracer methods usually seem indispensable.

Application of radioisotopes enables to eliminate measurement difficulties as the emission and detection of the tracer radiation does not depend on the physical conditions existing in the installation.

For dynamics investigations, the physical tracer is fully sufficient, since the only requirement concerned with its properties is to represent consistently and lastingly the labelled phase at all stages of the process. However in case when the kinetics or mechanism of the reaction undergoes investigation it is necessary to apply a chemical tracer, which takes part in all reactions and phase changes identically with the labelled substance. Observing a chemical likeness between the tracer and the labelled material is undoubtly the principle condition of the correct interpretation of the obtained results.

It also guarantees that the parameters of such processes as mass transfer, phase changes and material balance obtained on the grounds of radiometric measurements can be useful for technological estimation of the process.

In the reduction process of oxide slags in an electric furnace, the copper practically exists in two forms: as the metal and as the cuprous oxide, Cu_2O [1].

This fact makes the choice of a tracer almost evident, while copper irradiated in the thermal neutrons flux forms the isotope ^{64}Cu, the nuclear character of which predestinates it for radiometric measurements as the convenient and versatile tracer. ^{64}Cu emits γ-radiation of 511 keV; its half-life period amounts to 12·8 h, so it is sufficiently long lasting for the radiometric testing of all samples from one experiment while the residue of the isotope coming from the previous experiment doesn't cause distracting increase of the background.

The possibility of obtaining the ^{64}Cu isotope in form of metallic Cu and Cu_2O oxide made a basis for such planning the series of experiments that allowed to investigate thoroughly behaviour of the two copper forms in actual process.

Material for radiometric tests was collected during the entire process by means of sampling at the constant time intervals. During the final tapping samples were taken directly from the tapping spout.

The samples were then disintegrated, weighed and the activity of the tracer measured using specially designed measuring vessel. The measuring device was a scintillation detector of a NaI/Tl crystal type and a power supplier-counter system.

The reduction process of cuprous oxide in a slag takes place at the temperature of nearly 1300 °C in the course of reaction with powdered coke added as the reducer. Metallic copper settles gravitationally at the bottom of a furnace where it forms a layer of melted metal containing also the copper existing originally in metallic form in the processed slag [2, 3]. Application of a chemical tracer in the forms of ^{64}Cu or $^{64}Cu_2O$ aimed at determining the contribution of separate copper forms passing from slag to the final product.

The kinetics of the entire process of formation and sedimentation of the metallic copper includes three phenomena i.e.: chemical reduction of copper oxide, formation of metallic copper droplets and gravitational sedimentation of copper at the bottom of a furnace. Application of metallic ^{64}Cu and $^{64}Cu_2O$ respectively, was intended to separate these partial processes and to determine a contribution of each stage to the total rate of reduction process.

As complementary to radiometric tests, quantitative chemical analyses of samples were carried out to determine the total copper contents as well as the concentration of different copper compounds.

To obtain quantitative results also from radiometric tests the method of "standard" was applied. A small sample of irradiated material had been melted together with a certain mass of a slag in a laboratory furnace. Its activity was a function of the total activity of the tracer used, directly proportional to the mass ratio.

The standard prepared in such a way helped to obtain a mass-activity relationship and also to determine the entire activity of a tracer introduced into a furnace in one experiment. This enabled to complete a full material balance of copper, based upon the results of radiometric tests and chemical analyses.

Furthermore, metallographic observations were carried out. They aimed to determine the degree of dispersion of the metallic copper, the size of metallic concentrations etc.

3. Discussing the Results

Before interpreting and discussing the results obtained some simplifying assumptions were made. It has been assumed that the process run is fully consistent. Repeatability of the reduction process means that the final composition of separate phases after melting and tapping depends solely from the initial composition and time of the process. The reduction time was determined experimentally and remained constant in all experiments.

Another assumption made is that mixing was performed by a flow of gases formed during process, convection and gravitational sedimentation.

Thus, the dynamic parameters of the melt were not influenced by any external factor and the only possibility to stimulate the phase flow within the furnace was by changing the power-supply parameters.

Presented Fig. 1 is a good example of the results obtained. The curve illustrates changes of the pulse frequency in a series of samples, plotted against the process time. The tracer used in this experiment was active $^{64}Cu_2O$ introduced directly into the furnace, beneath the surface of the melt. Changes of the contents of tracer in samples refer to the period of tapping. Fig. 2 gives full information about

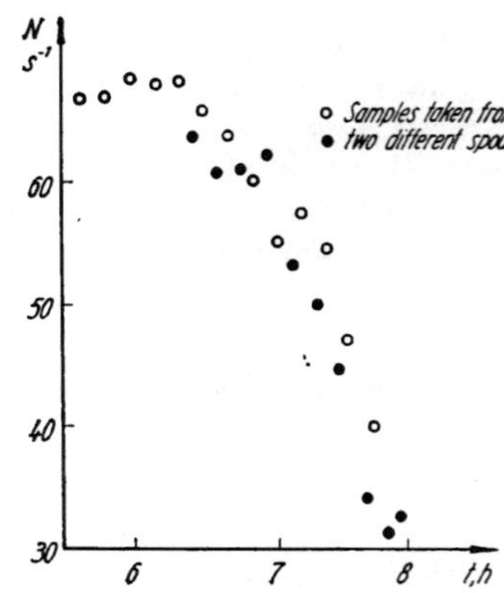

Fig. 1. Intensity of the tracer radiation in samples as a function of the process time

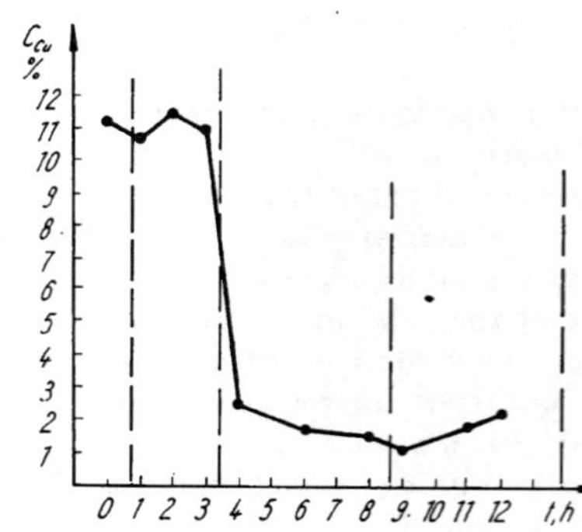

Fig. 2. Copper contents in slag samples collected during the course of the decoppering process

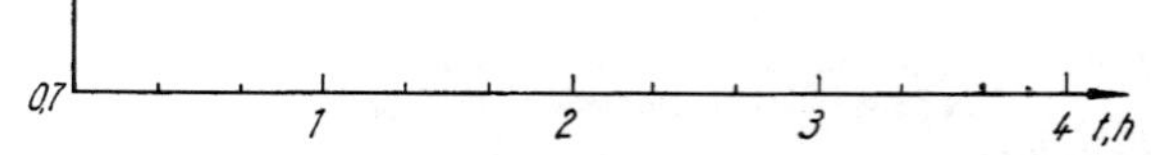

Fig. 3. Changes of the decoppering degree and the degree of reduction

the total copper concentration throughout the whole process.

It should be emphasized, however, that the slag decoppering occurs not only due to Cu_2O reduction but also due to sedimentation of metallic copper formed in the reduction process and that contained primarily in the slag.

The results of radiometric tests and quantitative analyses were used as a basis for further computations. Degree of reduction as well as degree of decoppering have been calculated for succesive experiments. In all experiments the degree of reduction values were considerably higher than those of decoppering degree, as shown in Fig. 3. The difference is pronounced at the early stage of the process, already. This proves the fact that the rate of the reduction process is relatively high, while sedimentation runs remarkably slower, thus being the limiting stage of the entire process.

However, sedimentation process runs differently for reduced metallic copper and for primary metallic copper in the slag. Cu_2O being reduced by direct contact with powdered coke and also with graphite electrodes forms fast growing nuclei of metallic phase, that, due to coalescence, form large drops settling at the bottom of a furnace [4, 5]. This is the cause of relatively low contents of reduction products in the waste-slag.

Small particles of primary metallic copper settle slowly, undergoing mixing processes preventing coalescence. Thus, the contents of original metallic copper in the waste slag is relatively high.

Therefore, from a technological point of view, time of reduction should not be exceeded beyond its necessary minimum, using full power supply. After reduction being finished, the next stage of the process should ensure favourable conditions for settling process.

This conclusion has also been confirmed by the results of material balance. The coefficients used in calculations, characterizing distribution of copper in each phase at separate stages of decoppering process, had been obtained by averaging corresponding values from different experiments.

Quantitative analyses were carried out not only to determine percentage of metallic and oxide copper but also in order to determine percent mass concentration of other elements, e.g. Pb, Fe.

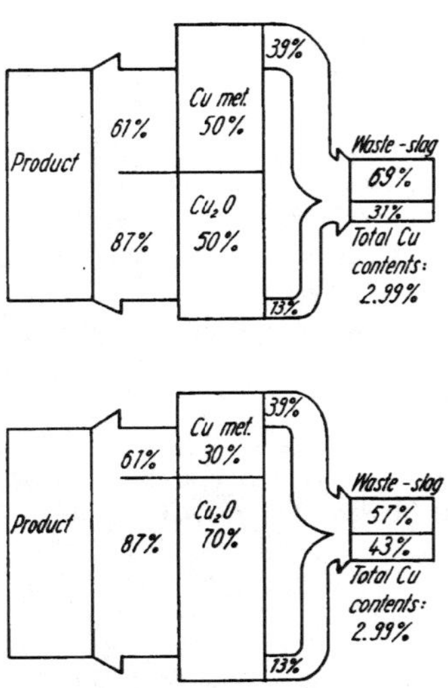

Fig. 4. Copper balance for the 12 % slag at the beginning of the reduction process

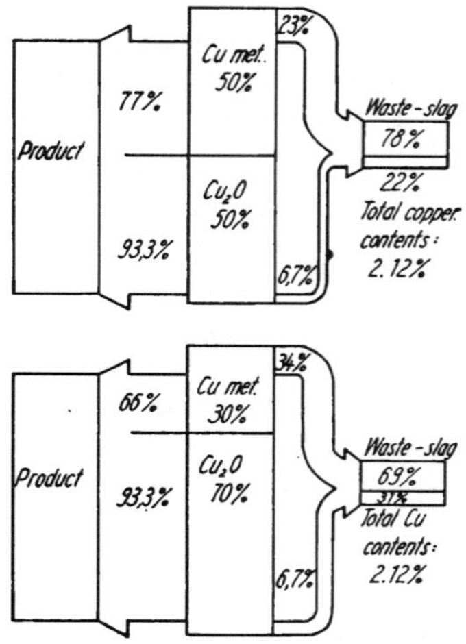

Fig. 5. Copper balance for the 12 % slag at the end of the reduction process

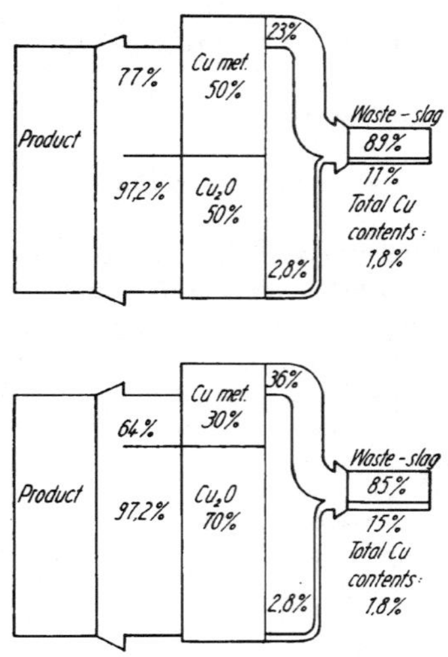

Fig. 6. Copper balance for the 12 % slag at the end of the final tapping

Material balance illustrates distribution of copper and explains the mechanism of its passing into products. Correspondingly to the increase of decoppering degree increases also the contribution of metallic copper in a waste-slag. (From 57%, through 69%, up to 85% at the end of tapping.) Copper balance is shown in the Fig. 4, 5, 6.

Therefore its impossible to improve general efficiency of the process by extending time of reduction. Exceeded reduction does not result in noticeable increase of decoppering degree, yet it requires increased expend of energy supplied.

On the contrary; forming advantageous conditions for sedimentation at the second stage of the process favours more complete decoppering and besides, results in energy saving effect.

The balance calculations seem to prove the previously made assumptions and confirm good accordance between proposed and observed mechanisms of decoppering processes.

4. Concluding remarks

- Application of the ^{64}Cu radioisotope in forms of metallic copper and cuprous oxide as a tracer for investigating the Cu_2O reduction process is fully effective method and can be succesfully used for quantitative interpretation.

- Different chemical forms of a tracer applied correspondingly to a labelled compound permit to investigate the mechanism of decoppering process and enable to differentiate behaviour of metallic and oxide copper.
- Method of "standard" enables to complete the quantitative balance of tracer's activity introduced to the process in succesive experiments.
- The activity of 37—100 GBq introduced in one portion of a tracer is fully sufficient for that sort of units being investigated, even if an experiment lasts around 24 h.
- Two main stages can be singled out in the decoppering process: the first is a period of Cu_2O reduction by coal and the second is metallic copper sedimentation.
- Primary contents of metallic copper in slag influences fundamentally the total amount of copper lost in a waste slag.

Received February 18, 1983

References

[1] *R. Chamer, K. Lemański, Cz. Mazanek,* Prace Naukowe Instytutu Chemii Nieorganicznej i Metalurgii Pierwiastków Rzadkich Politechniki Wrocławskiej 47 (1980) 11

[2] *J. Norwisz, R. Chamer, M. Staszweski, M. Wójtowicz* et al., Rudy i Metale Nieżelazne, 7 (1980) 291

[3] *T. Pomianek, L. Aadmkiewicz, St. Bratek,* Prace Instytutu Metali Nieżelaznych V. IX (1) (1980) 5

[4] *A. V. Vaniukhov, V. I. Zaytsev,* Slaky i Steyny Tsvetnoy Metallurgii, ed. by "Metallurgya", Moscow 1969

[5] *E. Iwanciw, A. Warczak, J. Wojewoda,* Rudy i Metale Niezelazne 9 (1977) 482

Isotopenpraxis **19** (1983) 11, S. 376 bis 378

Study of Volume Electric Charge in Radioactive Dielectrics by the Method of "Acoustic Sounding"*)

A. G. Rozno, V. V. Gromov

(Institute of Physical Chemistry of the Academy of Sciences of the USSR, Moscow)[1])

The formation of the volume electric charge in dielectrics containing radioactive β-isotope has been theoretically proposed. A method of calculating the value of the charge for both uniform and nonuniform distribution of the isotope in the volume is described. The method of the "acoustic sounding" has been used for the experimental study of charge distribution in the sample of crystalline $Y_3Al_5O_{12}$ (^{90}Y) and borosilicate glass ($^{90}Sr + {}^{90}Y$). The results obtained are in good accord with the theoretical evaluations. A conclusion has been drawn to the effect that the measurement of volume charge distribution can be used to study the absorption of β-radiation in a substance and spatial distribution of β-isotope in the material.

Die Bildung der elektrischen Raumladung in Dielektrika, die ein radioaktives β-Isotop enthalten, ist theoretisch ermittelt worden. Ein Verfahren zur Berechnung des Wertes der Ladung sowohl bei gleichmäßiger als auch bei ungleichmäßiger Verteilung der Isotope im Raum wird beschrieben. Das Verfahren der „akustischen Sondierung" wurde zur experimentellen Untersuchung der Ladungsverteilung in einer Probe kristallinen $Y_3Al_5O_{12}$(^{90}Y) und von Borosilikatglas ($^{90}Sr + {}^{90}Y$) eingesetzt. Die erzielten Resultate befinden sich in guter Übereinstimmung mit den theoretischen Bewertungen. Es wurde eine Schlußfolgerung dahingehend gezogen, daß die Messung der Raumladungsverteilung zur Untersuchung der Absorption von β-Strahlung in einem Stoff und der Raumverteilung eines β-Isotops in dem Material verwendet werden kann.

Keywords

dielectrics; electric fields; electric probes; radioactivity

1. Introduction

The electrification of β-radioactive substances was discovered back in 1900 [1]. However, the problem of the charge distribution in the volume of a substance has been considered just recently [2, 3]. It was shown that the charge is located in the subsurface layer, whose thickness is comparable to the mean length of the β-particle path. This layer is formed irrespective of the conductivity of the environment. The value of the charge is the difference of the number of particles leaving the volume element and the number of particles trapped in it. The relaxation process process mainly related to conductivity takes place along with the accumulation of the charge. Thus, the process of the charge formation can be described by the equation.

$$\frac{\partial \varrho}{\partial t} + \frac{\partial}{\partial x}(\sigma E) = f(x, t) \tag{1}$$

One-dimensional approximation facilitates the analysis and is in good accord with the experimental practice — during electric measurements flat parallel samples are usually used. The $f(x, t)$ function (the source function) describes the balance of trapped and escaped particles per one volume unit. σ is conductivity in a general case depending on the coordinate, E is the intensity of the electric field. In the case of low conductivity the relaxation process can be unessential up to the large values of the field and then

$$\varrho(x, t) = \int_0^t f(x, u)\, du \tag{2}$$

That is, the charge distribution corresponds to $f(x, t)$. The same correspondence is observed if σ does not depend on x

$$\varrho(x, t) = e^{-\frac{t}{\tau}} \int_0^t f(x, u)\, e^{\frac{u}{\tau}}\, du \tag{3}$$

here $\tau = \frac{\varepsilon\varepsilon_0}{\sigma}$ is the time of relaxation; ε is the dielectric constant of the material; $\varepsilon_0 = 8{\cdot}885 \times 10^{-12}$ F/m. In these

*) Presented at the 2nd Working Meeting "Radioisotope Application and Radiation Processing in Industry", 28th September to 1st October 1982, Leipzig

[1]) Address: Moscow, USSR, Leninski Prospect 31

cases the calculation of the ϱ distribution is brought down to the determination of $f(x, t)$.

2. Calculation of the source function

The digital calculation of the source function for samples of various dimentions and configuration containing uniformly distributed β-isotope was made on condition that singlevalued correspondence of the path and energy of β-particles [3] was observed. However, for one-dimensional case (flat parallel samples) one may use rather wide spread approximation, which gives a possibility to obtain results in the analytical form. It is wellknown [4] that the attenuation of the flux of β-particles in the substance accurately is described by the formula $N(x) = N_0 e^{-\mu x}$ where μ is the linear absorption coefficient. Then at the point of the material at a distance of x_0 away from the boundary ($x = 0$) the function of the source is calculated in the following way (Fig. 1):

$$f(x, t) = -\frac{q}{2} \cdot \frac{\partial}{\partial x}\left[\int_0^x A(u, t)\, e^{\mu(u-x)}\, du - \int_x^\infty A(u, t)\, e^{\mu(x-u)}\, du\right] \tag{4}$$

q — is the charge of the β-particle, $A(x, t)$ is the activity. In the case of a uniform distribution of the isotope we have:

$$f(x, t) = -\frac{q}{2} A(t)\, e^{-\mu x} \tag{5}$$

Then, for example, for β^--isotope ($q = -e$) at low conductivity of the dielectric ($t \ll \tau$) the charge distribution will be

$$\varrho(x, t) = \frac{e}{2} \exp(-\mu x) \cdot \int_0^t A(t)\, dt \tag{6}$$

The volume charge has a positive sign and it is located near the surface. The thickness of the layer is determined by the absorption coefficient.

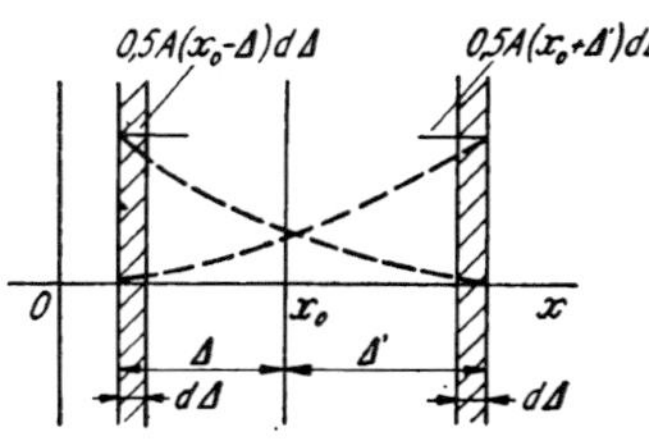

Fig. 1. The scheme to an evaluation of $f(x, t)$

Also of interest is the charging process in the case of nonuniform distribution of the isotope. Using (4) for points considerably ($x \gg 1/\mu$) removed from the boundary we obtain the following source function:

$$f(x, t) = -q \sum_{n=1}^{\infty} \frac{\partial^{2n} A(x, t)}{\partial x^{2n}}\, \mu^{-2n} \tag{7}$$

If the path-length of β-particles is short as compared to the size of non-uniformity

$$f(x, t) = -q \frac{\partial^2 A(x, t)}{\partial x^2}\, \mu^{-2} \tag{8}$$

Thus it can be seen that in the case of non-uniform distribution of the radioactive isotope in the volume of the dielectric (as well as on the boundary of the sample during uniform distribution) the volume charge can be formed.

3. Method of measuring the volume charge

The measurement of the volume charge distribution in radioactive dielectrics was made by the method of "acoustic sounding" [5, 6]. The principle of the method is presented in Fig. 2. In the process of measurements an acoustic wave is excited in the form of a narrow deformation band, which runs along the x axis. The current takes the path through the sample circuit. This current is recorded by the measuring system. The value of the current measured at a given moment is proportional to the charge density at the point of the sample, where the strain band is located.

$$I(t) \sim v_a \times S_{xx} \times \Delta \times \varrho(x)_{x=v_a t} \tag{9}$$

v_a — is the sound velocity, S_{xx} is the deformation,
Δ — is the width of the strain band. The signal is recorded

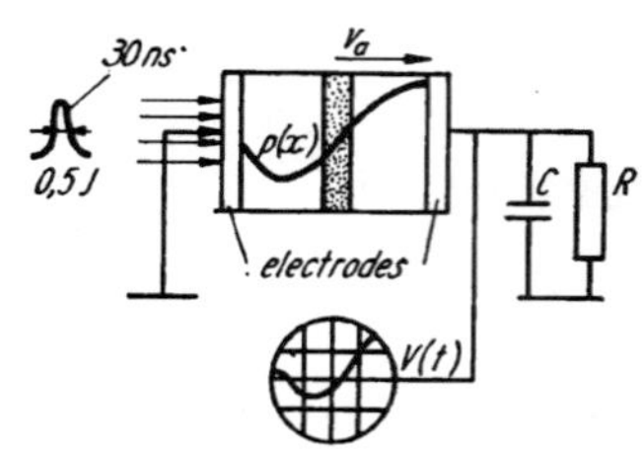

Fig. 2. The principal scheme of "acoustic sounding" method

oscillographically. Depending on the parameters of the recording circuit one can record the signal which reflects either the charge distribution or the distribution of the electric field in the sample [5].

For the implementation of the given scheme of measurement a rather narrow deformation band must be provided. For this, the surface of one of the electrodes applied to the sample, must be subject to the action of a short light pulse of high-power. Practically, a pulse with an energy of 1 J and duration of 30 ns causes in the sample the wave with the deformation amplitude $S_{xx} = 10^{-5}$ and extent $(0{\cdot}1 + 0{\cdot}5) \times 10^{-3}$ m. Obviously, the spatial resolution is determined by the width of the strain band. Charge sensitivity constituted -10^{-3} C/m³, field sensitivity $\sim 10^5$ V/m.

4. Volume charge in real radioactive dielectrics

For the experimental verification of the material presented above, we measured the volume charge in radioactive samples of single crystalline yttrium-aluminium garnet (YAG) and in borosilicate glass. In YAG the preset concentration of the ^{90}Y was provided by neutron irradiation. The radioactive preparation ^{90}Sr + ^{90}Y was introduced into the glass during melting. Fig. 3 shows the charge distri-

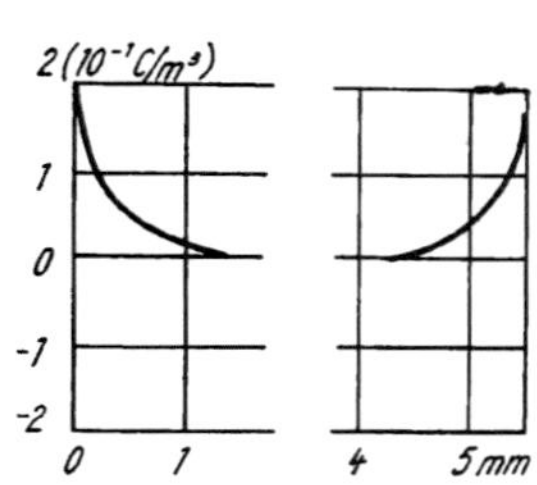

Fig. 3. The distribution of the electric charge in the sample of radioactive YAG

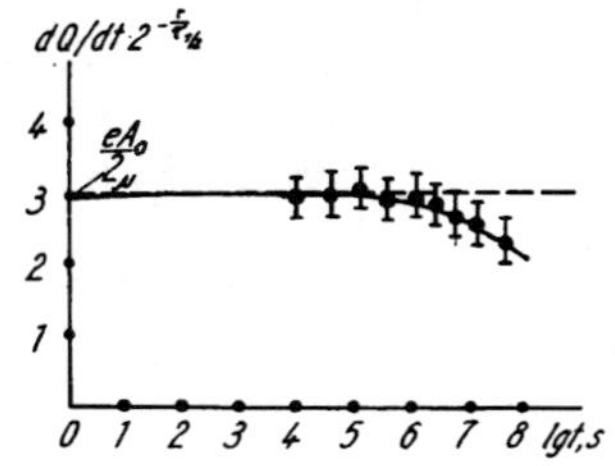

Fig. 4. The charge relaxation in YAG. The factor $2^{-\frac{t}{\tau_{1/2}}}$ is to compensate the decreasing of an activity

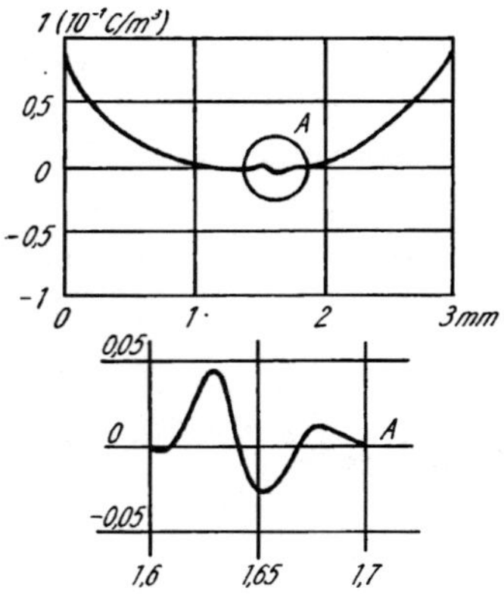

Fig. 5. The charge distribution in borosilicate glass containing ^{90}Sr + ^{90}Y

bution in the YAG sample 1 day after irradiation with a 5×10^{18} cm^{-2} beam of neutrons. As it was expected, the charge was located near the surface. Such distribution remained unchanged during the time of charge accumulation. Fig. 4 shows the dependence of the charge value on the time of accumulation. The relaxation time constitutes — 10^6 s. In ^{90}Sr + ^{90}Y activated sample glass, the maximum value of the volume charge and that of the electric field is achieved after 10^5 s. The corresponding distribution is shown in Fig. 5, its form is in accord with (6). The observed non-uniformity of the charge (Fig. 5), is, perhaps, due to the non-uniformity of the isotope distribution in the sample.

Conclusion

The results obtained support the assumption in [2, 3] and the evaluation presented in this work. Therefore, for high-ohmic dielectrics with a low mobility of the charge carrier (which is the reason for low radiation conductivity) one can study, measuring the distribution of the volume charge:
1) absorption coefficients of β-radiation in the dielectrics;
2) isotope distribution in the volume.

One can also estimate the activity value, necessary for such measurements. The parameter which characterizes the dielectric is the relaxation time. For the majority of dielectrics it lies within the limits $10^5 \div 10^8$ s. At sensitivity being 10^{-3} C/m^3 the minimal activity necessary for the measurements will constitute: $10^8 \div 10^{11}$ Bq/m^3.

$$A_{\min} = \frac{\varrho_{\min}}{e\tau_r} = 10^8 \div 10^{11}\,\mathrm{Bq/m^3}$$

The given evaluation shows that this method is rather sensitive to macro-non-uniform isotope distribution in the volume of the material. The thing of particular interest would be the study of radioactive isotope diffusion in dielectrics or diffusion of isotops to be activated by neutron irradiation.

Received February 18, 1983

References

[1] *P. Curie, M. Curie*, Compt. rend. **130** (1900) 647
[2] *V. V. Gromov*, Atomnaya energia **26** (1969) 250
[3] *A. G. Sakharov, V. V. Gromov, K. M. Merjanov*, Zhurnal fizicheskoi khimii **52** (1978) 433
[4] *A. A. Akhieser, V. B. Berestetsky*, Kvantovaya elektrodinamika, Gostekhteorizdat, Moskva 1953
[5] *A. G. Rozno, V. V. Gromov*, Pis'ma v JETF **5** (1979) 648
[6] *A. G. Rozno, V.V. Gromov*, Zhurnal fizicheskoi khimii **55** (1981) 1597

Isotopenpraxis 19 (1983) 11, S. 379 bis 383

On the Optimization of Concentration Measurements in Heterogeneous Materials Based on β-Backscatter Measurements*)

R. Burek, J. K. Zurawicz

(Technical University of Lublin, Poland)[1]

In the present paper it has been demonstrated that an error due to the variations of the chemical composition can be reduced by energy selection of the incident and backscattered β-radiation. The physical and practical aspects of this has been discussed.

In der vorliegenden Arbeit wird demonstriert, daß ein Fehler auf Grund von Veränderungen der chemischen Zusammensetzung verringert werden kann durch Energieselektion der einfallenden und der rückgestreuten β-Strahlung. Die physikalischen und praktischen Aspekte werden diskutiert.

Keywords

ashes; backscattering; beta particles; beta spectroscopy; chemical analysis; interfering elements; lead

1. Introduction

When the intensity of backscattered β-radiation is applied for radioanalytical purposes an error due to the variations of the chemical composition occurs. If we assume that the average backscattering coefficient of a compound depends upon the relative backscattering coefficients of the ingredients in the following way:

$$\bar{B} = \sum \alpha_i B_i$$

where α_i, B_i are the weight concentration and the backscattering coefficient, respectively, of the element i, then the influence of the variations of the chemical composition of the matrix is given by parameters defined as:

$$K_1 = \frac{\partial B}{\partial \alpha_i} : \frac{\partial B}{\partial \alpha_x} \tag{1}$$

where α_x — is the concentration of the element which has to be determined.

At the present time there is no simple way of avoiding such errors. In the presented paper it has been shown that such an error can be reduced by energy selection of the incident and backscattered radiation.

2. The relative energy distribution (RED) of the backward scattered radiation

According to the theory of the continuous diffusion, the RED of the backward scattered radiation may be approximated by the following function [1, p. 190]:

$$\varphi(u) = \frac{3}{4}(2u - u^2) \tag{2}$$

where $u = \frac{S}{S_D} = \frac{t}{t_D}$

S_D — the depth of the total diffusion

t_D — $\frac{S_D}{L}$ is the relative depth of diffusion

L — the average transport lenth

t, S — respective values for the depth of diffusion

α — the emerging angle of the backscatter radiation.

The relative energy of the backward scattered radiation is given by:

$$K = \frac{E}{E_0} = \left(1 - \frac{t}{t_{max}}\right)^{1.6}$$

where

$$t_{max} = \cos\alpha(1 - \cos\alpha)^{-1}$$

In the energy range from 0·5—2·0 MEV the relative diffusion depth t_p depends strongly on the atomic number and weakly on the primary energy of the particles [1].

When multielement samples are considered the average value of u is given by

$$\bar{u} = \sum_i \alpha_i \frac{t}{t_{D_i}}$$

This follows from the fact that t_D is evaluated from the transmission curves and corresponds to a certain depth, i.e. the half value layer.

The predictions of the RED as a function of the atomic number Z by means of Eq. (2) are in a good qualitative

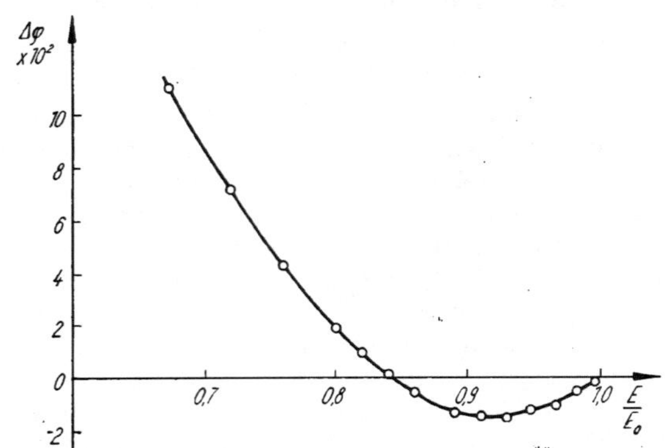

Fig. 1. Differences of the theoretical RED of β-particles backscattered through samples of different chemical composition (Cu + C, and equal atomic numbers)

---

*) Presented at the 2nd Working Meeting "Radioisotope Application and Radiation Processing in Industry", 28th September to 1st October 1982, Leipzig

[1]) Address: Department of Technical Physics, ul. Nadbystrzycka 38a Lublin/Poland

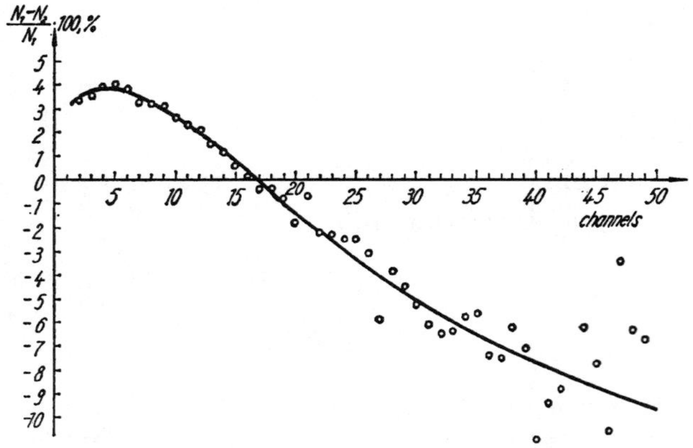

Fig. 2. Differences of the experimental RED of β-particles backscattered through samples of different chemical composition (SiO_2 and Fe_2O_3 + C) and equal atomic numbers (Z_{eff} = 10·8)

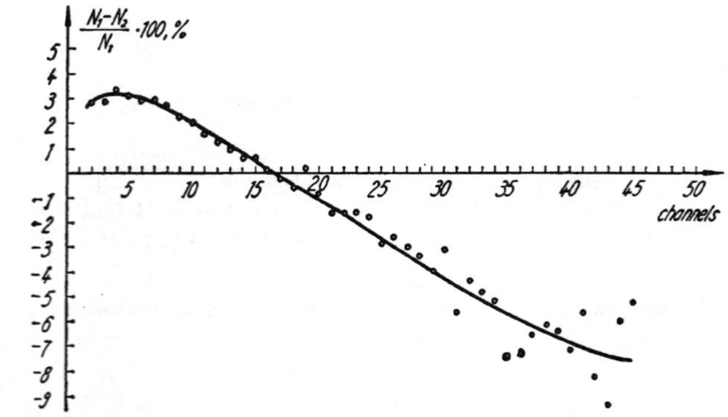

Fig. 3. Differences of the experimental RED of β-particles backscattered through samples of different chemical composition (SiO_2 + C and Fe_2O_3 + C), and equal atomic number (Z_{eff} = 9)

agreement with the experiments [1, p. 192]. Equation 2 allows to predict the influence of the chemical composition on the RED of the backward scattered intensity. Fig. 1 illustrates the differences $\Delta\varphi$ of the RED for mixture of Si + C and Cu + C, with equal effective atomic numbers.

The effect has been verifield experimentally. Fig. 2 shows the effect for SiO_2 and a mixture containing Fe_2O_3 + SiO_2 ($Z_{eff} = 10{\cdot}8$). Fig. 3 shows the effect for SiO_2 + C and Fe_2O_3 + C ($Z_{eff} = 9$).

As it is shown the experimental results agree qualitative with the theoretical predictions.

3. The dependence of the backscattering coefficient (BC) on the primary and backward scattered radiation

The primary energy of the β particles, has in a nonisotropic fixed geometry, a weak influence on the backscattering coefficient, particularly on its atomic number dependence.

The effect is caused by the absorption of the primary and scattered radiation. It is significant for targets of small atomic numbers and results in reduction of the BC with decreasing of the primary energy. However, when energy selection is applied the resulting BC(Z) curves are strongly influenced as on the primary as on the scattered energy.

Fig. 4, 5 and 6 show a series of such curves obtained with $^{90}Sr + {}^{90}Y{}^{204}Tl$ and ^{147}Pm sources. It is seen that backscattering coefficient increases with decreasing of the backscattered energy, systematically for ^{147}Pm and ^{204}Tl. For $^{90}Sr + {}^{90}Y$ the dependence is not monotonic. This indicates

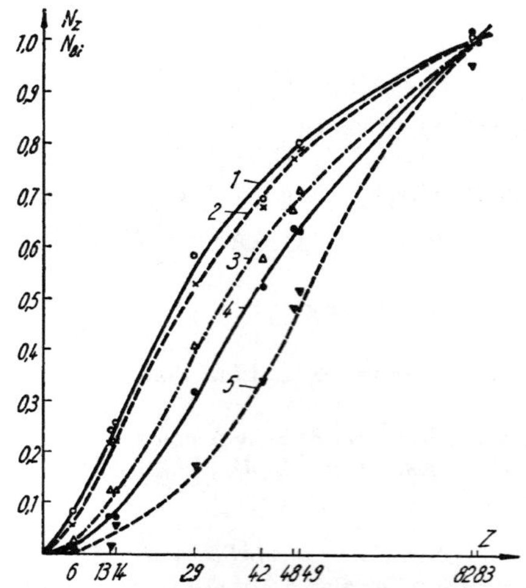

Fig. 4. Experimental curves of the dependence of relative intensity of backscattered β-particles on the atomic number

Source ^{147}Pm. The numbers 1, 2, ... 5 correspond to the following values of the discriminator bias: 0,5, 1,0 ... 4·5 V

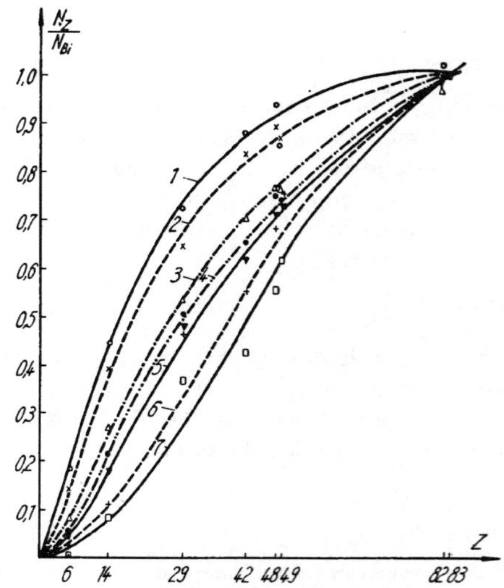

Fig. 5. Experimental curves of the dependence of relative intensity of backscattered β-particles on the atomic number

Source ^{204}Tl. The numbers 1, 2, ... 7 correspond to the following values of the discriminator bias: 0·5, 1·0 ... 4·5 V

that the ratio of eleastic to nonelastic scattered particles, varies in different parts of the spectrum of the secondary radiation.

4. The influence of the energy of primary and scattered radiation on the chemical matrix effect

The variety of the ratio of elastic to nonelastic scattered particles is resulting in a changeability of the relative values of the backscattering coefficients of heavy to light elements.

As shown in the Fig. 7 and 8 the relative backscattering coefficients $B(Z_1):B(Z_2)$ of SiO_2 and carbon to Fe_2O_3 depend strongly on the energy of the primary and backward scattered radiation.

This influences the coefficient of variation K, which is described by the eq. (1). The curves in Fig. 9 illustrate the dependences of K on the energy of the scattered radiation for a ^{147}Pm, $^{90}Sr + {}^{90}Y$ and ^{204}Tl source.

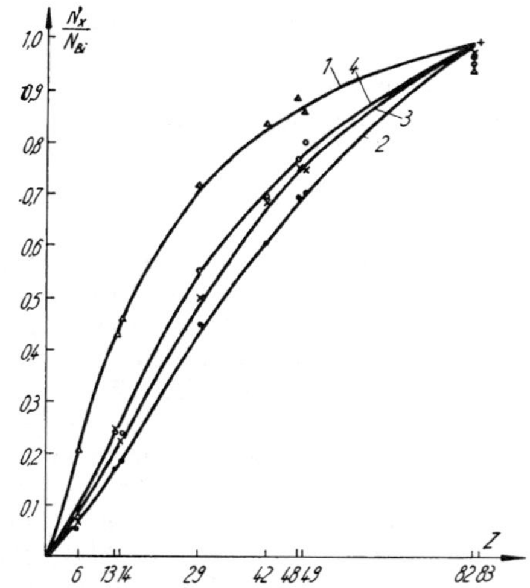

Fig. 6. Experimental curves of the dependence of relative intensity of backscattered β-particles on the atomic number

Source ^{90}Sr + ^{90}Y. The numbers 1, 2, ... 4, correspond to the following values of the discriminator bias: 0·5, 1·0 ... 4·5 V

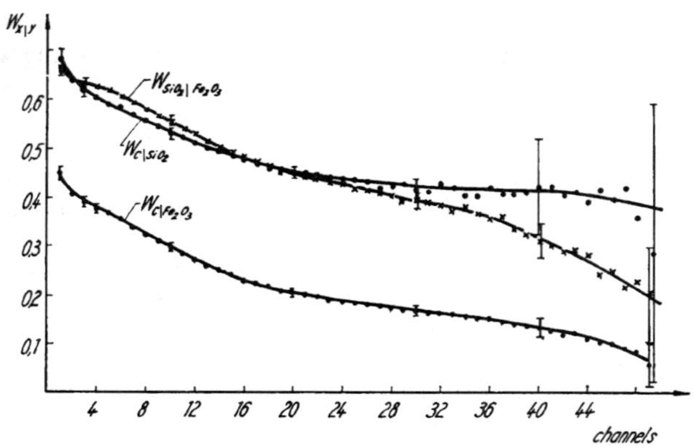

Fig. 7. The dependence of the relative backscattering coefficient on the energy of the backscattered radiation (source ^{204}Tl)

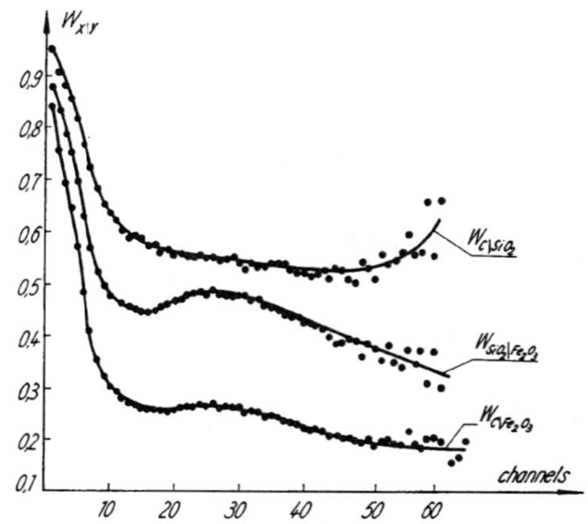

Fig. 8. The dependence of the relative backscattering coefficient on the energy the of backscattered radiation (souree ^{204}Tl)

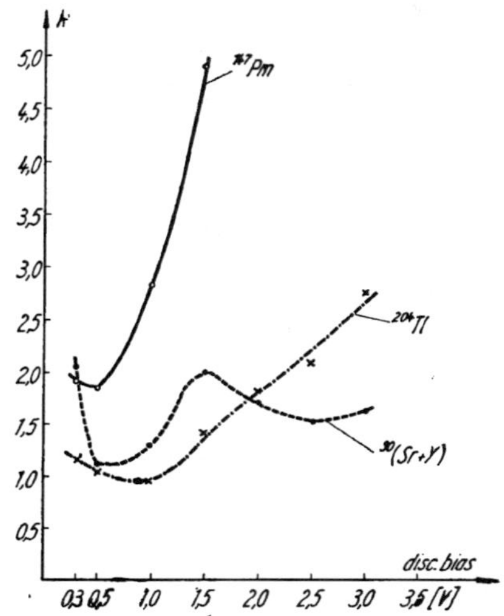

Fig. 9. Experimental curves of the dependence of the *K* factor on the energy of backscattered β-radiation

The dependences of *K* on the primary and secondary radiation indicates:

1) that a suitable choice of these energies may reduce the error, of the radiometric chemical analysis, due to variations of the chemical composition of the matrix.
2) that a suitable choice of n energies of the scattered radiation may afford independent determination of n + 1 elements.

5. Determinations of Pb in compounds

The calibrations measurements were performed on the following specimens: Pb, PbO, PbO_2, Pb_2O_3, $Pb(SCN)_2$, $Pb(NO_3)_2$, $PbSO_4$, with a source ^{85}Kr. All the samples were carefully pulverised and pressed into tablets.

Tab. 1. The Pb content normalized couting rate correlations

Backscattered energy (DISC. BIAS) mV	Regression line	Correlation coefficient	Standard error % Pb
3·5	$\alpha_{Pb} = 16{\cdot}95 + 2{\cdot}04\,N$	0·99673	1·237
2·3—3·3	$\alpha_{Pb} = 15{\cdot}78 + 114{\cdot}91\,N$	0·9984	0·8651
1·3—2·3	$\alpha_{Pb} = -54{\cdot}09 + 153{\cdot}2\,N$	0·9927	1·8526

Tab. 2. Composition of experimental and theoretical values of Pb for different energies of backscattered radiation

Compound	Experimentally determined Pb content			Theoretical
	Energy	(DISC. BIAS)	mV	value of Pb
	3·5—	2·3—3·3	1·3—2·3	
PbJ_2	83·32	85·85	85·17	44·94
$PbCl_2$	85·87	81·77	85·62	74·5
$PbBr_2$	75·41	77·21	77·94	56·4
$PbSO_4$	69·3	68·6	70·0	68·8
$Pb(SCN)_2$	65·43	64·88	66·6	64·1

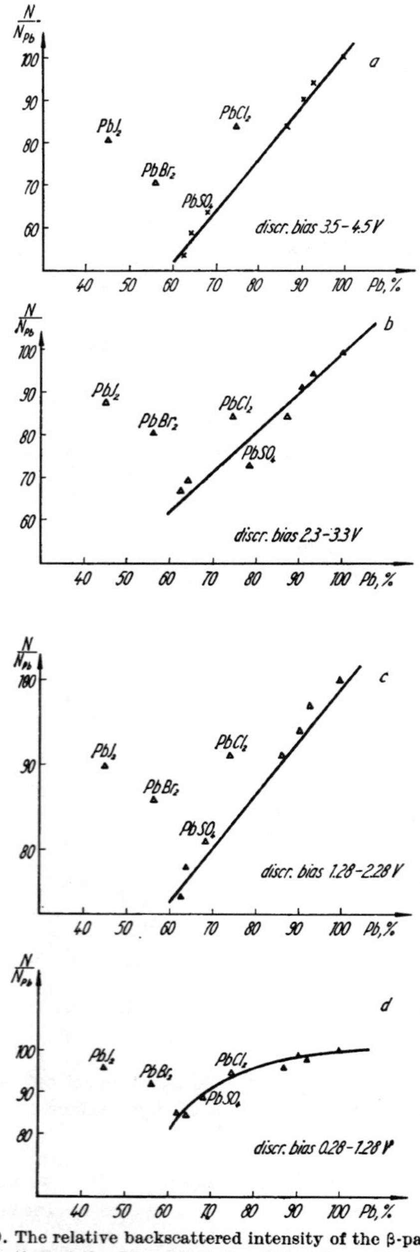

Fig. 10. The relative backscattered intensity of the β-particles as a function of the Pb content for different energies of the secondary β-particles.

curve a discriminator bias $3{\cdot}5 - \infty$ V; curve b discriminator bias $2{\cdot}3 - 3{\cdot}3$ V; curve c discriminator bias $1{\cdot}28 - 2{\cdot}28$ V; curve d discriminator bias $0{\cdot}28 - 1{\cdot}28$ V

Fig. 10 illustrates the calibration curves for the varies energies of the backscattered radiation. The equations of the best fit curves, their correlation coefficients and standard errors of estimation are listed in Tab. 1.

As it is seen the best fit of the calibration curve appears for the intermediate energy. Several other samples (PbJ_2, $PbCl_2$, $PbBr_2$) not used as standards have also been analysed. These compounds have been analysed to check the efficiency of the energy selection on the error of Pb determinations. As it is shown in Tab. 2 the energy selection results only on elements of low atomic numbers.

The difference between the experimentally determined weight fractions and the theoretical values listed in Tab. 2 are excessively large for $PbCl_2$, PbJ_2 and $PbBr_2$. In such a case energy selection has no influence on the reduction of the error of Pb determinations. For $PbSO_4$, $Pb(SCN)_2$ the energy selection results in reducing of the error.

6. Determination of combustible matter in float ash

The combustible matter contains carbon, oxygen and hydrogen. The highest contribution has carbon. The matrix consists of SiO_2, Al_2O_3, Fe_2O_3, CaO, MgO, Na_2O and K_2O.

For radioanalytical purposes float ash can be considered as a three element compound containing Fe_2O_3, SiO_2 and Carbon.

The measurements were carried out on 16 samples of float ash. The samples contained from 6 to 26% by weight of combustible matter and from 10 to 26% by weight of Fe_2O_3. The density of the float ash was 2000 ± 100 kg/m³. The bulk density of the samples ranged $380-900$ kg/m³, for a majority of cases the bulk density was 800 ± 100 kg/m³.

Some basic experiments were carried out with the source ^{204}Tl and ^{90}Sr + ^{90}Y to see the effect of the primary and scattered energy on the error due to the variations of the chemical composition.

The following factors has beet investigated:

1. The sensitivity of the backscattered intensity on changes of the combustible matter content:

$$S = \frac{1}{N_0} \frac{\Delta N}{\Delta \alpha_0}$$

where N_0 is the counting rate due a sample containing only inorganic matter
2. The correlation coefficient and standard error of estimation of the relationship between the intensity of the backscattered radiation and the content of combustible matter
3. The precision of the measurements defined as the standard deviation of 10 values of the backscattered intensity obtained dor freshly prepared samples. The precision includes the influence of the random agents, i.e. — bulk density, statistical fluctuation of the counting rate, inhomogenity.
4. The grain size effect coefficient for a sample containing 20% combustible matter, defined as:

$$H(d) = \frac{1}{N} \frac{\Delta N}{\Delta d}$$

where $N(O)$ — the couting rate for a pulverised sample
d — grain size, mm

To obtain the highest possible accuracy and most representative values for the investigated factors, the constans were obtained from a least square analysis. The results are tabulated in Tab. 3. As it is seen from the table the standards error of estimation decreases with decreasing of K. At the same time the grain size effect increases and the precision decreases. These indicate that the suitable choice of the primary and scattered radiation results in reducing of the error due to variations of the chemical composition. For a given source the parameter K may be applied for choosing of the backscattered energy.

7. Conclusions

When compounds contain leight elements ($Z < 30$) the error due to variations of the chemical composition may be

Kurznachrichten / Newlets

Strahlenrisiko:
WIE BEWERTET MAN DAS STRAHLENRISIKO

Obwohl der lebende Organismus ueber sehr wirksame Mechanismen zur Reparatur von Strahleneffekten in Zellen verfuegt, gibt es wahrscheinlich keine Dosisschwelle, bei deren Unterschreitung die Wahrscheinlichkeit fuer die Induktion bzw. Promotion von Krebs durch Strahlung absolut gleich Null ist. Die Wahrscheinlichkeit, das sogenannte Strahlenkrebsrisiko, nimmt jedoch mit sinkender Strahlendosis ab und ist daher bei niedrigen Dosen sehr klein. Die derzeitige mittlere Strahlenexposition pro Kopf unserer Bevoelkerung betraegt insgesamt etwa 300 mrem im Jahr, ausgedrueckt in Einheiten der effektiven Aequivalentdosis. Jeweils ca. 50% dieser Strahlendosis entfallen auf die Exposition durch natuerliche Strahlenquellen und auf die Anwendung ionisierender Strahlen in der Medizin. Demgegenueber werden weniger als 0,1% der gesamten Strahlenexposition durch die Emission radioaktiver Stoffe aus Kern- und Kohlekraftwerken verursacht. Im Strahlenschutz geht man von einer proportionalen Zunahme des Strahlenkrebsrisikos mit steigender Dosis aus. Die natuerliche Strahlenexposition (ca. 150 mrem/Jahr) haette demnach ein rechnerisches Strahlenkrebsrisiko zur Folge, das etwa - summiert ueber alle Tumorarten - 1% der derzeitigen beobachteten Krebshaeufigkeit entspricht. Fuer die radioaktiven Emissionen von Kraftwerken ergibt sich ein rechnerisches Strahlenkrebsrisiko in der Groessenordnung von etwa 0,001% der beobachteten Krebshaeufigkeit. Aus dieser Risikoanalyse ergibt sich als eine wesentliche Schlussfolgerung, dass nur ein sehr kleiner Bruchteil der beobachteten Krebsfaelle in der Bevoelkerung durch ionisierende Strahlen verursacht sein kann. Dieser kleine Bruchteil wird fast ausschliesslich durch die natuerliche und medizinische Strahlenexposition bedingt, waehrend der Beitrag infolge der Emission radioaktiver Stoffe aus Kraftwerken vernachlaessigbar ist. JACOBI betont, dass die zur Klaerung dieser Frage eingesetzten wissenschaftlichen Kommissionen der Vereinten Nationen (UNSCEAR) und der amerikanischen Akademie der Wissenschaften (BEIR-Committee) sowie die Internationale Strahlenschutzkommission (ICRP) in dieser Schlussfolgerung uebereinstimmen. (Gekuerzt nach W. JACOBI; Atomwirtschaft/Atomtechnik, Beilage Kernenergie und Umwelt, Juli-Heft 1983)

Kernenergie:
KERNKRAFT IN SCHWEDEN

Gemaess einer Mitteilung des Schwedischen Atomforums produzierten die Kernkraftwerke Schwedens im Jahr 1982 netto 37,3 Mia kWh. Dies entspricht 39% der gesamten Elektrizitaetserzeugung des Landes. Der Beitrag der Wasserkraftwerke machte 56% aus. Schweden steht damit nach Finnland und mit Frankreich ganz oben auf der Liste der punkto Atomstromanteil fuehrenden Laender der Welt. (SVA-Bulletin 5/83)

Strahlenexposition:
RADIOAKTIVITAETSUEBERWACHUNG IN DER SCHWEIZ SEIT 25 JAHREN

Im November 1982 hat die Eidg. Kommission zur Ueberwachung der Radioaktivitaet (KUER - c/o Physikal. Institut der Universitaet Perolles, CH 1700 Freiburg) einen 46seitigen Bericht herausgegeben, in dem einige Resultate der 25jaehrigen Messreihen enthalten sind. (Die KUER wurde am 16. Nov. 1956 gegruendet). Im Verlauf dieser Messzeit lagen die Maximalwerte der Radioaktivitaetspegel in den fruehen 60er Jahren (Kernwaffenversuche) und gehen seit 1963 stetig zurueck. Der Bericht kommt insgesamt zu der Feststellung, dass die Schweizer Bevoelkerung keine unzulaessigen Bestrahlungen erfahren hat. Trotzdem ist die Radioaktivitaet immer eine potentielle Gefahr, die staendig kontrolliert und ueberwacht werden muss. Die Spaltprodukte der chinesischen Wasserstoffbombe (1 ... 2 Mt) vom 27.6.1973 konnten in Luftproben (Aerosole) aus 12 500 m Hoehe am 17.7.1973 in der Schweiz mit einem Ge(Li)-Detektor nachgewiesen werden.

Strahlenexposition:
ANERKENNUNG VON KREBS ALS BERUFSKRANKHEIT

Die British Nuclear Fuels Ltd. (BNFL) hat sich mit ihrer Belegschaft und den zustaendigen Gewerkschaften ueber die Prozedur zur Anerkennung von eventuell durch berufliche Belastung ausgeloesten Krebsfaellen geeinigt. Diese Einigung laesst weiterhin die Moeglichkeit einer Anerkennung auf dem Prozesswege offen. In diesem Falle geht es jedoch nur um Anerkennung oder Abweisung der Klage, waehrend die betriebliche Vereinbarung vorsieht, auch geringere Strahlenbelastungen auf freiwilliger Basis im Falle einer Krebserkrankung als Ursache oder Mitursache gelten zu lassen. Die Hoehe einer Ausgleichszahlung wird von einem unabhaengigen Expertengremium festgelegt. (Atomwirtschaft/Atomtechnik - 4/83)

Endlagerung:
TON ALS MOEGLICHES WIRTSGESTEIN

Als "Schatzkammer alter Seidenstoffe" bezeichnen chinesische Archaeologen das kuerzlich entdeckte Grab einer Edelfrau aus dem 3. Jahrhundert v. u. Zeitr. in der Provinz Hubei. Aufsehen erregten neben verschiedenen anderen alten Textilien vor allem gut erhaltene, dekorative Seidenbaender, die an den Kleidern der vor 2300 Jahren Verstorbenen vorgefunden wurden. Unter dem Mikroskop stellten die Archaeologen fest, dass die Baender offenbar mit Nadeln gestrickt worden sind. Die Grabkammer liegt unter einer dicken Schicht von weissem Ton. Es wird angenommen, dass die Seidenstoffe der alten Chinesen dank der Schutzwirkung dieser dichten Schicht die Jahrtausende intakt ueberstanden haben. Ton ist ein moegliches Wirtsgestein fuer die Endlagerung radioaktiver Abfaelle. (SVA-Bulletin 8/83)

Strahlenrisiko:
FOLGEN DES BRANDES IM REAKTOR VON WINDSCALE/UK

Gemaess Aussagen des britischen Unterstaatssekretaers fuer Energie, John MOORE, gibt es keine Anhaltspunkte dafuer, dass irgendjemand als Folge des bekannten Brandes in einem militaerischen Bombenplutonium-Produktionsreaktor in Windscale vom 10./11. Oktober 1957 an Krebs erkrankt oder gestorben sei. MOORE sah sich am 22. Februar 1983 zu dieser Feststellung veranlasst, nachdem die Medien ueber eine Studie des National Radiological Protection Board (NRPB) in verschiedener Hinsicht missverstaendlich berichtet hatten. Im kuerzlich veroeffentlichten Bericht NRPB-R135 "An Assessment of the Radiological Impact of the Windscale Reactor Fire, October 1957" haben M.J. CRICK und G.S. LINSLEY rein rechnerisch die theoretisch moeglichen Folgen des Unfalls neu bewertet. Von der Ausnahme ausgehend, dass jede noch so kleine Strahlendosis mit einer gewissen Wahrscheinlichkeit gesundheitliche Wirkungen zeigt, berechnete das NRPB die Gesamtzahl von 260 moeglichen Schilddruesenerkrankungen. Aufgrund statistischer Erfahrungswerte koennten davon dreizehn toedlich verlaufen sein. Sieben weitere Personen waeren an anderen Krebserkrankungen verstorben. Laut NRPB sind die fuer die Berechnungen getroffenen Annahmen pessimistisch und geben einen theoretischen oberen Schaetzwert der Unfallfolgen. (SVA-Bulletin 5/83)

Endlagerung:
NOCH IMMER KEINE VERTRAGLICHE REGELUNG DER TIEFSEEVERSENKUNG

Eine Mehrheit der Vertragspartner des "Uebereinkommens ueber die Verhuetung der Meeresverschmutzung durch das Versenken von Abfaellen und anderen Stoffen" (Londoner Konvention) hat in ihrer 7. Konsultativsitzung vom 14.-18. Februar 1983 am Sitz der International Maritime Organization in London mit einer von Spanien eingebrachten Resolution dazu aufgefordert, die Tiefseeversenkung aller radioaktiven Stoffe auszusetzen, bis ein "Schlussbericht" ueber die technischen und wissenschaftlichen Grundlagen dieser Versenkung vorliegt. 19 Vertragsstaaten stimmten fuer die Resolution, sechs - naemlich Grossbritannien, Japan, die Niederlande, die Schweiz, Suedafrika und die USA - stimmten dagegen. Fuenf Staaten, darunter die BRD, Frankreich und die Sowjetunion, enthielten sich der Stimme. Die Resolution erreichte damit die fuer die Verbindlichkeitserklaerung von Beschluessen ueber die Anlagen des Uebereinkommens noetige Zweidrittelmehrheit nicht. Verschiedene Laender, die in den letzten Jahren radioaktive Abfaelle in der Tiefsee versenkt haben, wollen dem Aufruf der Londoner Konvention nicht folgen. So hat Grossbritannien angekuendigt, dass es die fuer dieses Jahr vorgesehene Versenkung von 4200 t Abfall planmaessig durchfuehren werde. Auch die Niederlande, welche noch Ende 1982 zugunsten der Beseitigung ihrer radioaktiven Abfaelle an Land auf die Tiefseeversenkung verzichten wollten, werden gemaess einer Verlautbarung ihrer Verhandlungsdelegation mindestens 1983 sich noch an Versenkungsaktionen beteiligen, da bisher kein anderer Lagerplatz gefunden wurde. Belgien, das der Londoner Konvention nicht angehoert, hat keine Aenderung seiner Versenkungspolitik angekuendigt. Der Schweiz wuerde der sofortige Verzicht auf die Tiefseeversenkung aehnliche Probleme wie den Niederlanden bringen. (SVA-Bulletin 5/83)

Strahlenexposition:
EINE ALTERNATIVE ZUR ERHOEHUNG DES RADONGEHALTS DER ATEMLUFT DURCH DAS ABDICHTEN VON WOHNRAEUMEN

Eine naheliegende und einfache Moeglichkeit, wirksam Heizenergie einzusparen, besteht im besseren Abdichten von Fugen und Spalten, durch die sonst kostbare Warmluft aus Wohnraeumen entweicht. Als Nebenwirkung des Abdichtens erhoeht sich aber u.a. der Radon-Gehalt in der Wohnraum-Atemluft, woraus sich nach heutiger Kenntnis ein zusaetzliches Lungenkrebsrisiko ergibt. Das radioaktive Edelgas Radon stammt aus dem Zerfall von Stoffen wie Uran, Thorium und Radium, die von Natur aus im Erdboden und in Baumaterialien vorhanden sind. In einer vergleichenden Untersuchung haben W. Burkart vom Eidg. Institut fuer Reaktorforschung (EIR) und S. Chakraborty von der Hauptabteilung fuer die Sicherheit der Kernanlagen (HSK) kuerzlich mit quantitativen Abschaetzungen unterstrichen, dass das Energiesparen durch Abdichten um Groessenordnungen riskanter ist als die Produktion der jeweils gleichen Energiemenge aus Kohle, Oel, Gas oder Uran. So waere bei der Kernenergie unter Einschluss des gesamten Brennstoffkreislaufs aufgrund theoretischer Abschaetzungen insgesamt mit 0,08 bis 0,21 hypothetischen Todesfaellen durch Unfaelle und Krankheiten pro Mia kWh zu rechnen. Die nach den gleichen Kriterien theoretisch errechnete Zahl hypothetischer Todesfaelle durch die zusaetzliche Radon-Belastung pro Mia kWh Heizenergie, die durch einfaches Abdichten eingespart wuerde, geben BURKART und CHAKRABORTY mit 51,22 an. Damit waere das Risiko mehrere hundert Mal kleiner, wenn statt abzudichten mit Atomstrom geheizt wuerde. Die Autoren weisen deutlich darauf hin, dass das Radon nur ein Teilaspekt der zusaetzlichen Innenluftverschmutzung durch Abdichtmassnahmen ist. Neben Radon koennen in schlecht beluefteten Raeumen auch Schadstoffe wie Formaldehyd aus Isolierschaeumen und Stickoxyde aus Gasoefen zu gesundheitlichen Problemen fuehren, wenn nicht besondere Kontrollmassnahmen getroffen werden. (SVA-Bulletin 5/83)

Ankündigungen / Announcements

JOSEPHSON-EFFEKT: PHYSIK UND ANWENDUNGEN
Isd-wo Mir, Moskau 1984, 672 S. (Russ.)

SYSTEMTHEORIE VIELER TEILCHEN
Basarow, I.P.; Nikolajew, P.N.
Isd-wo Moskauer Universitaet, 1984, 320 S. (Russ.)

MATHEMATISCHE METHODEN DER QUANTENPHYSIK
Isd-wo Mir, Moskau 1984, 448 S. (Russ.)

ELEKTRONENSPEKTROSKOPIE DER KRISTALLE
Nemoschkalenko, W.W.; Aleschin, W.G.
Isd-wo Nauk. dumka, Kiew 1983, 400 S., 2. Aufl. (Russ.)

N e u e Z e i t s c h r i f t:

"ISOTOPE GEOSCIENCE"

Vol. 1 (No. 1) March 1983
Multilingual journal, largely English
Elsevier Scientific Publishing Comp., Amsterdam
Herausgeber (Editor-in-Chief): Faure, G.
Erscheinungsweise: quartalsweise

Aus dem Inhalt:

Ar-40/Ar-39 incremental heating plateaus for biotites with excess argon
Foland, K.A.

The origin of salts on Mt. Erebus and along the coast of Ross Island, Antarctica
Jones, L.M., et al.

Stable-isotope geochemistry of Hercynian granitoid rocks from the Sarrabus Massif, Sardinia
Brotzu, P., et al.

Sr-isotope geochemistry of fluorite mineralization, Mexico
Kesler, S.E., et al.

U and Th decay-series disequilibria in historical lavas from the Eolian Islands, Tyrrhenian Sea
Cepaldi, G., et al.

Veröffentlichungen / Publications

TRACER KINETICS AND PHYSIOLOGIC MODELING - PROCEEDINGS OF A SEMINAR
Springer-Verlag, Berlin [West] 1983, 509 pp.
Lambrecht, R.M. (ed.)

ION IMPLANTATION: EQUIPMENT AND TECHNIQUES - PROCEEDINGS OF THE 4TH INTERNATIONAL CONFERENCE
Ryssel, H.; Glawischnig, H. (eds.)
Springer-Verlag, Berlin [West] 1983, 580 pp.

GEOCHEMISTRY OF GROUNDWATER

Back, W.; Letolle, R. (eds.)
Elsevier Scientific Publishing Company, Amsterdam and New York, 1981, 370 pp.

Among other papers:

S- and O-isotopes as tracers for the origin of sulfate in Lake Creteil(Southeast of Paris, France)
Chesterikoff, A., et al.

The Madrid Basin aquifer: preliminary isotopic reconnaissance
Lopez Vera, F., et al.

Radiocarbon dating in groundwater of the aquifer confined in the Lower Triassic sandstones of the Lorraine region (France)
Blavoux, B.; Olive, P.

Uranium isotopes and Ra-226 content in the deep groundwaters of the Tri-State region(USA)
Cowart, J.B.

Carbonate geochemistry of vadose water recharging limestone aquifers
Thrailkill, J.; Robl, T.L.

A geochemical method of determining dispersivity in regional groundwater systems
Wood, W.W.

STAATLICHES AMT FUER ATOMSICHERHEIT UND STRAHLENSCHUTZ DER DDR:

SAAS - 304

Untersuchungen von Neutronen- und Gammafeldern von Tokamak-Fusionsreaktoren Investigation of neutron and gamma fields of tokamak fusion reactors - 38 S.
Doerschel, B.

SAAS - 305

Untersuchungen zur Wechselwirkung von verfestigten radioaktiven Abfaellen mit Steinsalz durch Hygroskopizitaet - Interaction between solidified radioactive wastes and rock salt due to hygroscopicity - 32 S.
Koerner, W.

SAAS - 306

Modelluntersuchungen zur Waermeuebertragung von abgebrannten Brennstoffkassetten in gasgefuellten Transportbehaeltern - Model investigations of the heat transfer from spent fuel assemblies in gas-filled skipping casks - 30 S.
Nitsche, Fr.

UNITED KINGDOM ATOMIC ENERGY AUTHORITY:

AERE-R 10628

The development of an EMAT system for the accurate measurement of the phase velocity of shear waves - October, 1982, 8 pp.
Allen, D.R., et al.

AERE-R 10642

An automatically calibrated universal measuring equipment for noise in semiconductors - December, 1982, 25 pp.
Cox, C.E.

AERE-R 10713

The TGSL finite element subroutine library - December, 1982, 29 pp.
Jackson, C.P.

AERE-R 10819

Potential for the application of nuclear techniques in selected developing countries in Africa - February, 1983, 139 pp.
Clayton, C.G.

CLM-R 230

Ripple transport in INTOR - 1982, 12 pp.
Ashby, D.E.T.F., et al.

CLM-R 234

Models and codes for the plasma edge region - December, 1982, 121 pp.
Harbour, P.J., et al.

AERE-R 10734

The elimination of small voids in UO_2 during irradiation - October, 1982, 19 pp.
Small, G.J.

AERE-M 3280

Inspection of mineral insulated thermocouple closure and junction welds by metallography - November, 1982, 9 pp.
Broomfield, G.H.

AERE-R 10692

The use of laser anemometry to measure swirl velocity in a firing production diesel engine December, 1982, 33 pp.
Wigley, G., et al.

ND-R 777 (R)

An engineering design study for storage and disposal of intermediate level waste - November, 1982, 132 pp.
Griffin, J.R., et al.

Ausgewählte Beiträge /
Recently published papers

ATOMWIRTSCHAFT-ATOMTECHNIK 28(1982)6:

Verfuegbarkeit der Kernkraftwerke in der Welt im Jahre 1982
Szeless, A.; Oszuszky, F.

Jodfreisetzung nach Stoerfaellen im Aufloeserabgassystem
Obrowsky, W.; Nagel, K.

Verbrauch von Primaerenergie nach Energietraegern 1970-1981

Kernenergie in der Weltelektrizitaetserzeugung 1981

ANGEWANDTE CHEMIE 95(1983)4:

Extrem weitreichende D-Isotopeneffekte auf chemische Verschiebungen in C-13-NMR-Spektren von Verbindungen mit konjugierten Doppelbindungen
Berger, St.; Kuenzer, H.

Isotopeneffekte bei der Niederdruck-, Hochdruck- und Duennschichtchromatographie der Lanthanoiden
Hufnagel, A.; Specker, H.

ATOMKERNENERGIE-KERNTECHNIK 42(1983)3:

Der Elektronenlinearbeschleuniger, eine vielseitige Strahlungsquelle
Kneissl, U.

Experimente zur Grundzustands-Protonenradioaktivitaet
Hofmann, S.

Medizinische Therapie mit pi^--Mesonen
Walder, E.

Beschleunigernuklide fuer die Nuklearmedizin
Pfeiffer, G.

Eine eingehende Studie ueber die Ausbeute der Cerenkov-Strahlung in transparenten Medien
Swailem, F.M.; Moussa, M.B.

Ein theoretisches Modell fuer die Neutronenstreuung an Benzol
Sefidvash, F.

Ein Beitrag zur Frage des Transfers von Technetium 99 in die Milch
Wiechen, W. u.a.

ATOMNAJA ENERGIJA 54(1983)2:

Measurement of Np-237 (n,2n) Reaction Cross Section at 14.8 MeV Neutron Incident Energy (Russ.)
Gromova, E.A., et al.

The Effect of Deuterium-Ion Bombardment on the Structure of Polycrystallic Niobium (Russ.)
Pisarev, A.A., et al.

Research Neutron Fields at the IR-100 Reactor (Russ.)
Barabanov, S.A., et al.

Comparative Neutron-Physical Calculations for Fast Reactors (Russ.)
Yordanov, Yo.D., et al.

A Universal Neutron Irradiator Based on Californium-252 (Russ.)
Bak, M.A. et al.

Investigation of the Effect of Hydrogen-Ion Bombardment on the Structure and Chemical Composition of High-Nickel Alloy (Russ.)
Guseva, M.I., et al.

Neutron Activity of the Earth and the Cl-Ar Neutrino Experiment (Russ.)
Barabanov, I.R., et al.

V.G. Khlopin Radium Institute Ion Synchrotron Complex (Russ.)
Perfilov, N.A., et al.

ATOMNAJA ENERGIJA 55(1983)1:

The Water-Chemical Regime of a Fast-Reactor Power Complex (Russ.)
Musikhin, R.N., et al.

Pore Migration and Redistribution of Actinides in (U,Pu) O_2 at the Beginning of Fuel-Element Operation (Russ.)
Bakhteyev, A.N., et al.

Fast-Neutron Fission Cross-Section of Cf-249 (Russ.)
Kupriyanov, V.M., et al.

Multilevel Parametrization of Pu-239 Total and Fission Cross Section in thd Resonance Region of Neutron Energies (Russ.)
Bakalov, T., et al.

Detection of Radionuclides Using an Automatic Training Classifier (Russ.)
Sazykin, B.V.

Optimization of Heterogenetic Shielding Compositions Based on SSM (Russ.)
Nikitin, N.V., et al.

A Medical Apparatus Based on an Air-Cored Betatron (Russ.)
Amelichkin, V.I., et al.

Effect of the Efficiency of Registering Fission Events on the Measured Value of the Average Number of Prompt Fission Neutrons (Russ.)
Malinovsky, V.V., et al.

On the Gamma Equivalent of Ionizing Radiation Sources (Russ.)
Bochkaryov, V.V., et al.

The Determination of Light-Rock-forming Elements by Means of Activation Analysis Using a Laser Generator of Neutrons (Russ.)
Varik, V.I., et al.

ATOMNAJA ENERGIJA 55(1983)2:

Assessment of the Cumulative Ejection of Cesium-137 and its Propagation Over the Region of an Atomic Power Station (Russ.)
Silantyev, A.N., et al.

Measuring the Characteristics of U-238 and U-236 Spontaneous Fission (Russ.)
Belenki, S.I., et al.

Concentration Field Structure of Artificial Radionuclides in the Baltic and North Seas in the Spring of 1981 (Russ.)
Styro, D.B., et al.

Isotopic Equilibrium in Hydrogen-Hydride Systems of Intermetallic Compounds (Russ.)
Andreyev, B.M., et al.

A Method of Evaluating the Resolving Power of Gamma-Ray Well Logging (Russ.)
Pritchin, B.P.

An Installation for Automatic Measurement of Water Radioactivity (Russ.)
Surkov, Yu.A., et al.

The Determination of Light Rock-Forming Elements by Means of Activation Analysis Using a Laser Generator of Neutrons (Russ.)
Varik, V.I., et al.

Stopping Cross-Section of $N\text{-}14^+$ Ions in Al at Energies up to 30 keV/Nucleon (Russ.)
Yalyshko, S.V.

ATOMNAJA ENERGIJA 55(1983)3:

Activation Methods of Determining Oxygen, Nitrogen and Fluorine in Nuclear Fuel (Russ.)
Ovechkin, V.V., et al.

A Method of Calculating the Optimal Number of Points for Monitoring Local and Global Radioactive Pollution of the Environment (Russ.)
Makhonko, K.P.

The Determination Sensitivities of Certain Elements with $Z < 42$ Using a Deuteron Activation Method in a Cyclotron (Russ.)
Vakilova, G., et al.

Estimating Systematic Errors in Calculating Criticality by the Monte Carlo Method (Russ.)
Zolotukhin, V.G., et al.

An Automated System of Monitoring the Technological Process of Isolating Transplutonium Elements (Russ.)
Bikineyev, V.A., et al.

Eliminating Tracks on the Surface of Alkaline Glass (Russ.)
Zelenski, V.F., et al.

Tab. 3.

Source	Energy of Back-scattered	Coefficient of variations K	Sensitivity $\frac{1}{N_0}\frac{dN}{dA}$	Correlation coefficient	Standard error of estimation	Precision of measure-ment % float ash	Grain size effect coeff. $\frac{1}{N(O)}\frac{\Delta N}{\Delta d}$
^{204}Tl	Total	1·5 K 2·0	−0·0050	−0·837	2·9	1·7	−0·047
^{90}Sr + ^{90}Y	Total	2·0 K 3·0	−0·0040	0·727	3·9	0·6	−0·021
^{204}Tl	Energy selec.	1·45	−0·009	−0·981	1·25		
		2·25	−0·012	−0·979	1·30		
		3·50	−0·013	−0·903	1·99		
^{90}Sr		1·80	−0·007	−0·8697	3·16		
		2·45	−0·009	−0·8309	3·57		
		3·50	−0·012	−0·6624	4·80		

reduced by energy selection of the incident and scattered radiation. The effect is caused by the change-ability of the ratio of simple scattered to multiscattered betas, particulary by diffusion. This results in the dependence of the RED of the backscattered intensity, for samples of equal atomic numbers and different chemical composition. For a given source of primary energy the backscattered energy may be choosen by means of the coefficient of variations K.

Received February 26, 1983

References

[1] *H. W. Thummel*, Durchgang von Elektronen und Betastrahlung durch Materieschichten. Berlin: Akademie-Verlag 1974

Isotopenpraxis **19** (1983) 11, S. 383 bis 386

Meßanordnung zur Untersuchung der Radiothermolumineszenz von Polymeren

R. Kunze, E. Linke

(Akademie der Wissenschaften der DDR, Zentralinstitut für physikalische Chemie)[1])

In der Arbeit wird eine Anordnung zur Messung der Radiothermolumineszenz an dünnen Polymerschichten, die bei der Temperatur von $T = 80$ K bestrahlt werden, beschrieben. Auf den Aufbau der für die Bestrahlung verwendeten Elektronenstrahlquelle und der Kühl- und Heizeinrichtung wird näher eingegangen.
Als Elektronenquelle wird ein Elektrodensystem für Oszillographenröhren verwendet, das Stromdichten von $I \geq 10^{-6}$ A/cm² bei variabler Elektronenenergie von 5 keV bis 25 keV liefert. Besondere Vorteile zeigt die Anordnung durch die räumliche Einheit von Bestrahlungsquelle, Probenhalterung und SEV, so daß Messungen der isothermen Lumineszenz bei tiefen Temperaturen unmittelbar nach Bestrahlungsende möglich sind.

In the present work, an appliance for the measurement of the radiothermoluminescence of thin polymer layers irradiated at a temperature $T = 80$ K is described. The design of the electron beam source used for the irradiation as well as of both the cooling and the heating systems is presented in detail.
An electrode system of oscillograph tubes is used as an electron source yielding flux density $I \geq 10^{-6}$ A/cm² with electron energy being variable from 5 keV to 25 keV. The appliance offers particular advantage by the aggregation of irradiation source, sample holder and SEV, rendering possible measurement of isothermal luminescence at low temperature immediately after irradiation.

Keywords

electron beams; luminescence; measuring instruments; polyacrylates; polymers; thermoluminescence

1. Einleitung

Radiothermolumineszenzuntersuchungen (RTL) an bei tiefen Temperaturen bestrahlten hochmolekularen Verbindungen fanden vor 15 Jahren Eingang in die physikalisch-chemische Grundlagenforschung an Polymeren [1, 2].

Sie sind eine sehr empfindliche Methode zum Nachweis von ionischen Teilchen in organischen Festkörpern und können zur Aufklärung von strahlenchemischen Elementarprozessen eingesetzt werden. Im einzelnen liefert die RTL Aussagen über

- molekulare Beweglichkeiten von Kettensegmenten
- Strukturübergänge im Polymer
- Elektronenhaftstellen und
- den Einfluß von Additiven und physikalischer bzw. chemischer Struktur des Polymers auf den Ladungseinfang bzw. auf die Diffusion der Ladung.

Die hier aufgebaute Meßeinrichtung soll speziell für Untersuchungen an E- und R-Resists [3, 4] auf der Basis von

[1]) Anschrift: DDR-1199 Berlin-Adlershof, Rudower Chaussee 5

Polymethylmethacrylat (PMMA) eingesetzt werden. Sie kann aber auch für RTL-Messungen an Ionenkristallen und anderen Festkörpern bei tiefen Temperaturen verwendet werden.

Da bei allen geplanten Untersuchungen auch der Abklingmechanismus der isothermen Lumineszenz unmittelbar nach Bestrahlungsende untersucht werden soll, wurde bei der Entwicklung der Meßeinrichtung besonderer Wert auf die Möglichkeit gelegt, die Bestrahlung und Messung an ein und demselben Ort vorzunehmen, d. h. ohne Transport der Probe.

2. Aufbau der Meßeinrichtung

2.1. *Mechanischer Aufbau*

Die Meßeinrichtung setzt sich aus folgenden Hauptbaugruppen zusammen:

- Hochvakuumpumpstand mit Rezipienten
- SEV mit Verstärker und Registriereinrichtung
- Elektronenstrahlquelle mit Spannungsversorgung
- Probehalterung mit Kühl- und Heizeinrichtung

In Abb. 1 ist der prinzipielle Aufbau der Meßeinrichtung mit den einzelnen Baugruppen dargestellt.

Auf die beiden letzteren Baugruppen, die im Rahmen der vorliegenden Arbeit entwickelt und aufgebaut wurden, wird im folgenden näher eingegangen.

2.2. *Elektronenstrahlquelle*

Aus den Versuchsparametern ergaben sich an die Elektronenstrahlquelle folgende Forderungen:

- Strahlstrom: $I_S = 10^{-6}$ A ... 10^{-9} A
- Strahlbündeldurchmesser: $d_B \leqq 3$ mm
- Strahlspannung: $U_B = 5$ kV ... 25 kV
- Strahlablenkung in x- und y-Richtung um $\pm 1,5$ cm
- hohe Lebensdauer der Katode und widerstandsfähig gegen Belüftungen der Anlage

Besonders die beiden letzten Forderungen stellten eine Einschränkung für den Aufbau der Elektronenstrahlquelle dar. Durch Verwendung eines Elektrodensystems einer Oszillographenröhre wurde der Versuch unternommen, die oben gestellten Anforderungen zu erfüllen. Es wurden Systeme des Typs B 13S6 und B 7S4 eingesetzt, die uns vom VEB Funkwerk Erfurt freundlicherweise zur Verfügung gestellt wurden.

Die in den Oszillographenröhren eingesetzten Oxidkatoden wurden durch eine Wolframhaarnadelkatode ersetzt, die sich durch eine wesentlich höhere Widerstands-

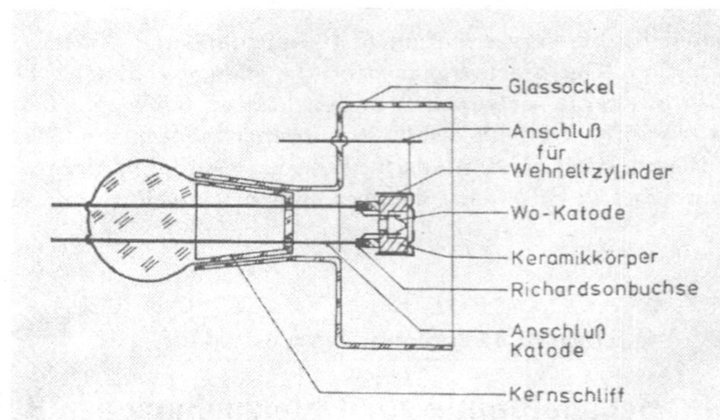

Abb. 2. Schematische Darstellung des Aufbaus der Katode
(Halterung und Anschlüsse für die übrigen Elektroden der Elektronenstrahlquelle wurden aus Gründen der Übersichtlichkeit nicht eingezeichnet)

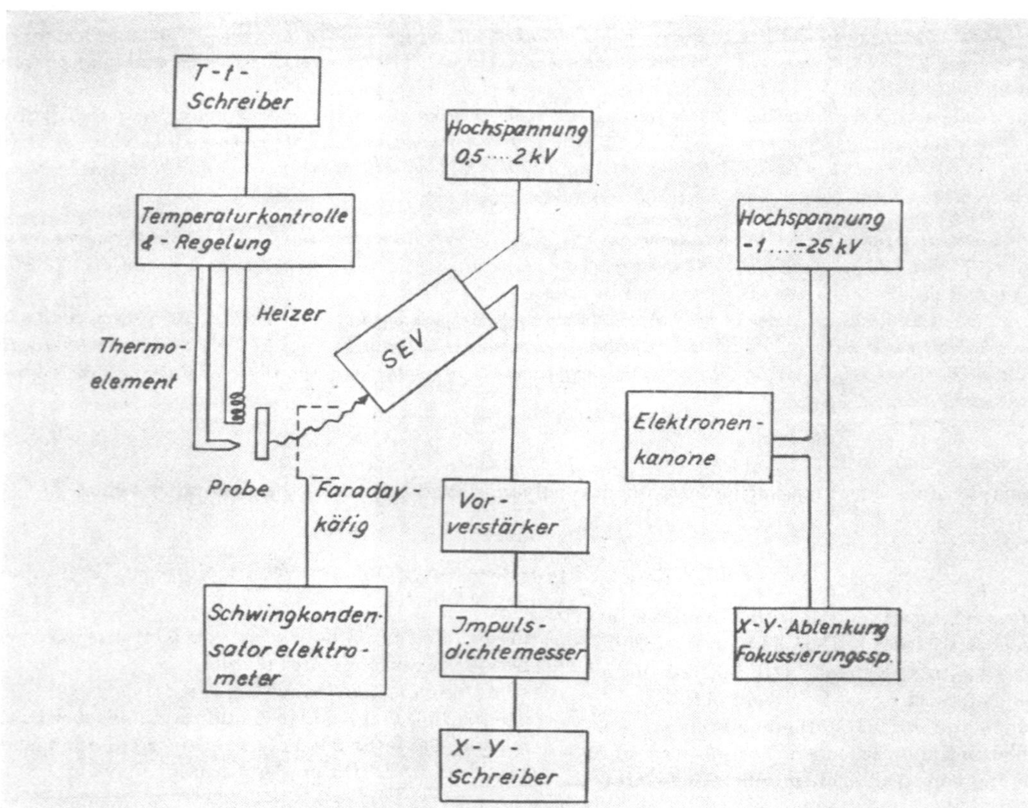

Abb. 1. Prinzipaufbau der Meßeinrichtung

fähigkeit gegenüber Belüftungen des Rezipienten [4, 5] auszeichnet. Der Sockel wurde ebenfalls umgestaltet, um ein schnelles Auswechseln der Katode zu ermöglichen. Die Katodenhalterung wird von einem Kernschliff mit zwei Nicosildrahtdurchführungen gebildet, auf die ein Keramikkörper mit der Katode über Buchsen aufgesteckt wird. Der Keramikkörper sichert die Zentrierung und einen reproduzierenden Abstand der Katode von der Lochblende des *Wehnelt*zylinders (Abb.2).

Zur Gewährleistung der hohen Beschleunigungsspannung von $U_B = 25$ kV wurde das Prinzip der Nachbeschleunigung [6] verwendet, da das Elektrodensystem der Oszillographenröhre selbst nur bis ~ 8 kV spannungsfest ist. Zur Nachbeschleunigung wurde nach der Ablenkeinheit noch eine weitere Elektrode, die durch ein feines Kupfernetz realisiert wird, angebracht. Da die Nachbeschleunigung erst nach dem Ablenksystem erfolgt, kann mit relativ kleinen Ablenkspannungen gearbeitet werden.

Das gesamte Elektrodensystem befindet sich in einem teilweise doppelwandigen Glaskolben, der an dem Vakuumflansch über einen Kovarring angeschweißt wurde und hochspannungsfest gegenüber dem Rezipienten ist. Abb. 3 zeigt die vollständige Elektronenstrahlquelle mit dem Anschlußflansch für den Rezipienten.

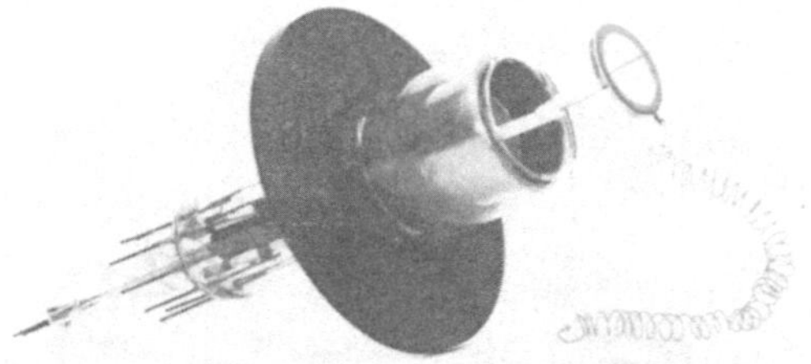

Abb. 3. Elektronenstrahlquelle mit Anschlußflansch und Nachbeschleunigungselektrode (vorn rechts)

Die Hochspannungsversorgung der Elektronenstrahlquelle erfolgt über einen Gleichspannungsregler Typ 4212 (VEB Startron Fürstenwalde) mit einem Spannungsbereich von -1 kV bis -25 kV. Die Quelle wurde mit dem negativen Pol der Hochspannung verbunden, damit die zu bestrahlenden Proben auf Erdpotential liegen.

Die einzelnen Versorgungsspannungen für den *Wehnelt*zylinder, die 1. Beschleunigungsstufe, die Fokussierungselektrode und die Nachbeschleunigung werden über eine Potentialteilerschaltung erzeugt. Die Ablenkungsspannungen für x- bzw. y-Richtung sind im Bereich von jeweils 0 bis ± 110 V variierbar. Zur Messung des Strahlstromes, des Strahlquerschnittes und zur Kontrolle der Strahllage dient ein *Faraday*käfig, der mit verschiedenen Formen von Eintrittsblenden ausgerüstet werden kann. Der Käfig kann mit einer Genauigkeit von $\pm 0{,}05$ mm in der Ebene senkrecht zur Strahlrichtung über den Elektronenstrahl gefahren werden. In Abb. 4 sind zwei Strahlungsprofile für $U_B = 10$ kV und 15 kV, die mit dem *Faraday*käfig mit einer einfachen Schneide als Eintrittsblende aufgenommen wurden, dargestellt. Da sich der *Faraday*käfig unmittelbar vor der Probenhalterung befindet, kann bei Verwendung einer Lochblende die Strahllage auf der bestrahlten Probe festgestellt werden, bzw. mit Hilfe der x-y-Ablenkung des Elektronenstrahles eine durch die Lochblende vorgegebene Stelle der Probe bestrahlt werden. Zusätzlich läßt sich noch eine

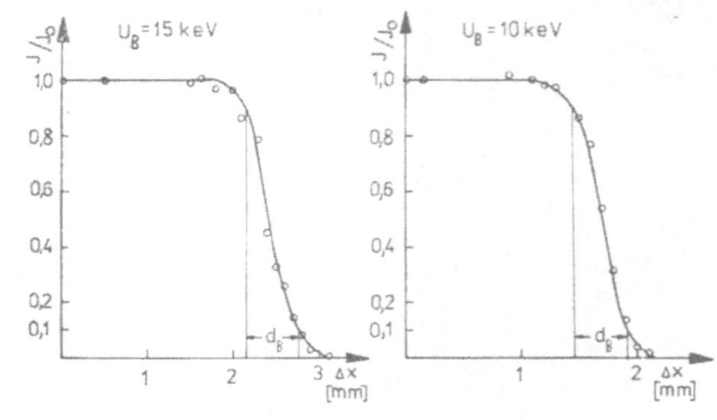

Abb. 4. Experimentell ermitteltes Strahlprofil mit einer einfachen Schneide für zwei Beschleunigungsspannungen U_B

(Als Strahlquerschnitt d_B wird der Abstand zwischen der Ordinate für 90 % und 10 % des totalen Wertes definiert

optische Justierung des Elektronenstrahles mit Hilfe eines kleinen ZnS-Schirmes durchführen, der an den Ort der Probe gebracht wird.

Durch Variation der Fokussierungsspannung läßt sich die Brennfleckform in gewissen Grenzen variieren, z. B. bei Verwendung des Systems B 13S6 von einer rechteckigen Form mit den Abmessungen von 3×1 mm² über die minimalen Abmessungen von $0{,}7 \times 0{,}6$ mm² zu einer rechteckigen Form von $1{,}0 \times 3{,}5$ mm² mit ansteigender Fokussierungsspannung. Die damit erzielten Stromdichten für die Bestrahlung der Proben liegen im Bereich $\geqq 10^{-6}$ A/cm². Zur Erreichung von Ladungsdichten auf der Oberfläche von 10^{-6} bis 10^{-4} C/cm², wie sie für Elektronenstrahl-Resists charakteristisch sind, werden damit Bestrahlungszeiten im Bereich von wenigen Sekunden benötigt.

2.3. *Heiz- und Kühleinrichtung*

Die Probenhalterung soll die Möglichkeit bieten, die bestrahlten Proben bei Untersuchung der thermisch stimulierten Lumineszenz von 80 K bis zu etwa 500 K mit linearen Heizraten von 2 K/min bis zu 25 K/min zu erwärmen. Dabei wurde besonderes Augenmerk auf einen guten Wärmekontakt zwischen Probe und Probenhalterung und auf ein schnelles Abkühlen auf die Temperatur des flüssigen Stickstoffes gelegt. Die hier gewählte Variante ist in Abb. 5 dargestellt. Zur Kühlung der Probe wird die zylinderförmige Halterung (Durchmesser: 42 mm) von flüssigem Stickstoff durchflossen. Die vordere metallische Deckplatte unter der die zu bestrahlende Probe angebracht wird, gewährleistet eine gleichmäßige Temperaturverteilung über der 12×12 mm² großen Probe.

Zur Reduzierung des Wärmeaustausches der Probenhalterung mit dem Rezipienten wurden die Durchführungen für den Stickstoffdurchfluß und für die Heizpatrone doppelwandig ausgeführt. Nach einer Kühlzeit von ungefähr 6 min wird in der Probe die Endtemperatur von 80 K erreicht. Durch den Anschluß eines Vorratsbehälters mit flüssigem Stickstoff kann diese Temperatur über einen längeren Zeitraum aufrechterhalten werden. Die Aufheizung der Probe erfolgt über eine 40 W Heizpatrone, die in die Halterung eingeschoben wird. Bei der Aufheizung wird die Zufuhr von flüssigem Stickstoff unterbrochen.

Durch Variation und Regelung der Heizleistung werden die geforderten linearen Heizraten im vorgegebenen Temperaturbereich von 80 bis 500 K realisiert. Durch Gegen-

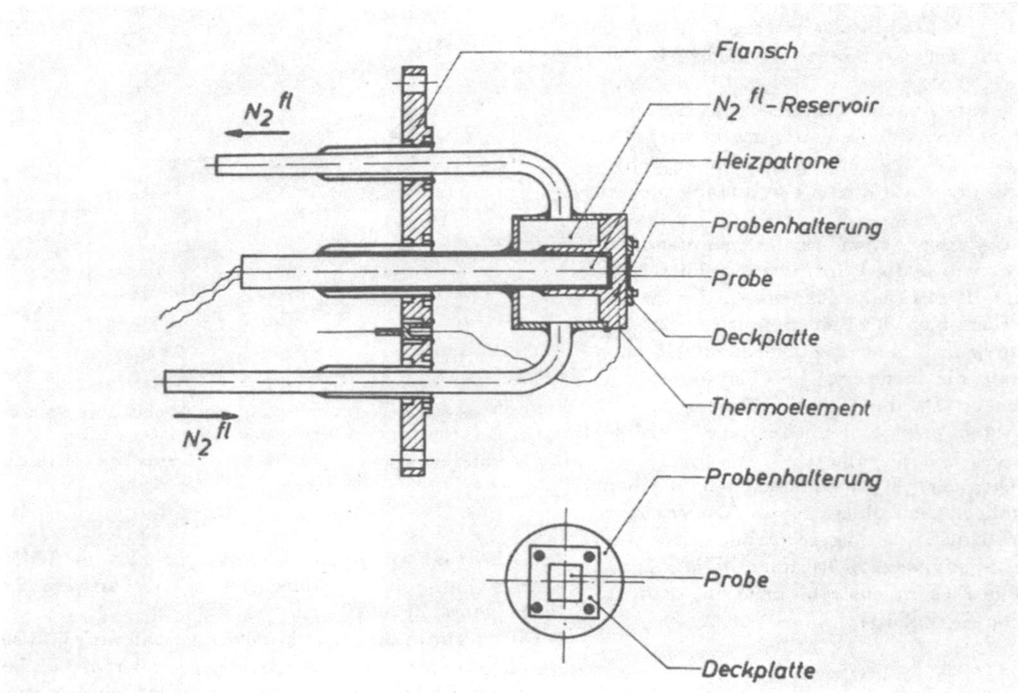

Abb. 5. Prinzipskizze der Heiz- und Kühleinrichtung für die RTL-Untersuchungen

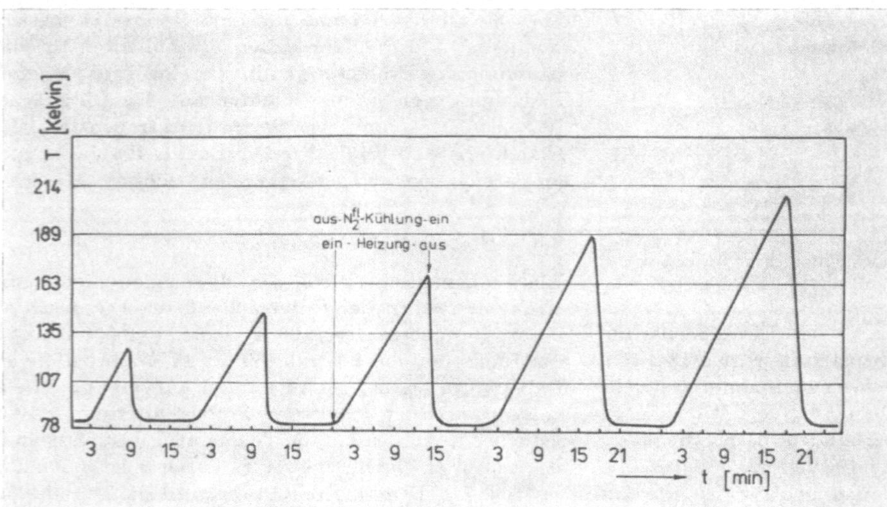

Abb. 6. Temperaturverlauf in der Probe beim partiellen Aufheizen bzw. Kühlen

kühlen mit flüssigem Stickstoff bei gleichzeitigem Heizen läßt sich außerdem eine gewünschte Temperatur fest einstellen, so daß die Möglichkeit besteht, auch bei den entsprechenden Peaktemperaturen isotherme Abklingkurven aufzunehmen. Weiterhin können durch ein spezielles Temperaturregime Thermolumineszenzkurven nach der Methode des schrittweisen Aufheizens der bestrahlten Proben aufgezeichnet werden und damit eine verbesserte Auflösung der einzelnen Glowpeaks erreicht werden.

In Abb. 6 ist der mit einem Fe-Ko-Thermoelement aufgenommene Temperaturverlauf an der Probe für eine partielle Aufheizung dargestellt.

Über erste Ergebnisse von RTL-Messungen an bestrahlten Polymethylmethacrylat (PMMA) und Polyethylen wird in einer folgenden Arbeit berichtet.

Eingegangen am 1. 3. 1983

Literatur

[1] *M. Dole* (Ed.), The Radiation Chemistry of Macromolecules Vol. I. New York: Academic Press 1972

[2] *P. Bräunlich* (Ed.), Thermally Stimulated Relaxation in Solids. Berlin – Heidelberg – New York: Springer Verlag 1979

[3] *R. F. W. Pease*, Contemp. Phys. Vol. 22 (1981) 265

[4] *W. P. Derkatsch, W. M. Korsinskij*, Mikroelektronika 11 (1982) 291

[5] *H. Göllnitz* (Hrsg.), Vakuumelektronik. Berlin: Akademie-Verlag 1978

[5a] *G. A. Haas, J. F. Jensen*, Vakuum-Technik 9 (1960) 76

[6] *E. Gundert, W. Scheffernicht*, Elektronische Rundschau 11 (1957) 107

Isotopenpraxis 19 (1983) 11, S. 387 bis 390

On the Possibilities of the Direct Gamma-Spectrometry in Natural Waters*)

B. Manouchev, T. Boschkova, L. Tsankov, V. Gurev, I. Kojucharov, G. Grosev

(Sofia University "Klement Okhridsky", Faculty of Physics, Dep. of Atomic Physics)[1])

In this paper some problems are considered, related to the application of direct gamma-spectrometry in natural waters: absolute sensitivity calibration of the detector system and its theoretical evaluation by means of a spherical model of the detector; possibility for a mathematical stabilization of the experimental calibration using numerical filtering methods; evaluation of the sensitivity of the method is also performed. Several spectra are presented of the gamma-field in some natural basins in Bulgaria.

In dieser Arbeit werden einige Probleme der Anwendung der direkten Gamma-Spektrometrie in natürlichen Gewässern untersucht: die absolute Empfindlichkeitseichung des Detektorsystems und seine theoretische Bewertung mit Hilfe eines sphärischen Detektormodells; die Möglichkeit einer mathematischen Stabilisierung der Versuchseichung unter Verwendung numerischer Filterverfahren; Bewertung der Empfindlichkeit des Verfahrens. Mehrere Spektren des Gamma-Feldes werden für einige natürliche Gewässer Bulgariens vorgestellt.

Keywords

calibration; gamma detection; gamma spectroscopy; environment; mathematical models; water

1. Introduction

We known about 20 papers dealing with different problems connected with the direct measurements of the gamma-field in natural waters. The first of them date from the late 50-ies and early 60-ies. The poor resolution of the scintillation counters, used in that time did not permit sure identification of the artifical radionuclides on a background due to the natural nuclides whose spectrum is rich with gamma lines. This brought to a wide development and application of the radiochemical separation of the nuclides, despite its obvious unconveniences: the necessity of time consuming chemical operations which require skilled personnel, do not permit automatization and continuous control.

The development of the gamma spectrometric equipment (especially the high-efficiency and high resolution semiconductor detectors and compact multichannel analyzers permitting automatic processing of the acumulated gamma spectra) makes possible the wide introduction of the direct gamma spectrometry for the monitoring of the radiation purity of natural waters and the waters used for cooling in the nuclear power plants.

The direct gamma spectrometry can be accomplished in two forms: either by spectrometry of water specimen (diverted in bypass mode from the main water flow, or taken periodically) in a low background shield or by submerging the detector unto the investigated water volume. Both modes permit an automatic and continuous (or semicontinuous) checking and have their advantages and disadvantages. The first variant proposes a more convenient arrangement of the detector system, while the second one does not require a low background shield which is much more complex than in the case of spectrometry of solid samples. The considerably lower background in the case of a submerged detector (if the dimensions of the water volume are sufficient) combined with the larger amount of water being permit a higher sensitivity. On the other hand, in this case difficulties arise with the detector arrangement in a water-tight container and assuring of low temperature necessary for the operation of a semiconductor detector.

It seems, that for practical applications up to now the first variant is preffered (for ex. [1]). But it is obvious that considerable theoretical and experimental comparisons (including numerical estimates) are necessary in order to choose the variant, more effective for mass applications.

In this paper we will concentrate on some principal problems connected with the direct gamma-spectrometry of large volumes of natural waters with a detector, submerged in them: determination of the absolute sensitivity (experimental and theoretical) of the detector system, an estimate of the sensitivity and the precision of the method, mathematical stabilization of the energy calibration.

2. Determination of the absolute sensitivity of the detector

Experimentally the absolute sensitivity can be determined by the measurement the count rate from a detector, submerged in a sufficiently large volume of water with known specific activity. We, however, propose a method which permits the experimental photosensitivity calibration to be done used a point source of radiation. We measure the count rate of the photo-peak I_{Ph}^{R} when a source of activity A is placed at a distance R from the detector. The source and the detector, both mounted in water-tight containers are submerged in a sufficiently large volume of water. Then the photosensitivity P_{exp} in the approximation of a sphericaly symmetric detector is given by the expression:

$$P_{exp} = \frac{4\pi R^2 I_{Ph}^{R} e^{\mu_{H_2O} R}}{A \mu_{H_2O}} \quad (1)$$

*) Presented at the 2nd Working Meeting "Radioisotope Application and Radiation Processing in Industry", 28th September to 1st October 1982, Leipzig; supported in part by the International Atomic Energy Agency

[1]) Address: 5 Anton Ivanov blvd., BG-1126 Sofia, Bulgaria

Here μ_{H_2O} is the linear absorption coefficient for gamma rays in water for a given energy E.

In the real case the sensitivity diagram of the detector has not a spherical symmetry. For this purpose the procedure, described above should be applied for different directions and an average for the results should be taken. An additional correction is implied by the asymmetry of the water-tight container due to which the water layer at a constant distance source detector is not the same in the different directions.

It is expendient the experimentaly measured sensitivity, at different energies to be fitted to the theoretically calculated. The theoretical calculation of the sensitivity or the efficiency of the gamma-ray detectors, especially in the case of distributed radiation sources is connected with considerable mathematical difficulties. The explicit calculation of the full sensitivity for a system, composed of a cylindrical source and cylindrical detector leads to a non-resolvable analytically 5-fold integral [2]. Another approach to solving of this problem is the Monte Carlo method [3, 4], which require too long calculations.

For this reason an attempt was made for theoretical estimation of the sensitivity of a cylindrical detector by substituting of the latter with a sphere of equal volume. We consider a point source at a distance ϱ from the center of a spherical detector of radius T_0. The detector can be surrounded by a spherical container having radius R_1 and wall thickness d_2. Let μ_0, μ_1 and μ_2 be the corresponding full attenuation coefficients for the radiation (all of them functions of energy) for the material of the detector, the medium between the source and the detector and the container. The full sensitivity of the detector can be expressed as:

$$T_s = t_1 t_2 t_3 t_4 \tag{2}$$

where:

$t_1 = (1 - \sqrt{1 - k_0^2})/2$ is the probability for emission of a gamma quantum toward the detector ($k_0 = R_0/\varrho$);

$t_2 = \exp(-\mu_4 \overline{L_1})$ is the probability for the gamma quantum, emitted toward the detector to reach the latter without being absorbed in the medium between the source and the detector. $\overline{L_1}$ is the averaged geometrical path of the photon from the source to the detector, which is given by

$$\overline{L}_1(K_0K_1) = R_1\left[K_0 - E\left(\arcsin\frac{R_0}{R_1}, K_1\right) + (1 - K_1^2)F\left(\arcsin\frac{R_0}{R^1}, K_1\right)\right] \Big/ (K_1 \arcsin K_0) \tag{3}$$

where $k_1 = R_1/\varrho$ and $F(x, y)$ and $E(x, y)$ are the uncomplete eliptic integrals of the first and the second kind, respectively.

t_3 is the probability for the gamma quantum to pass the container without any interaction. If the container is not too thick we can admit $t_3 = \exp(-\mu_2 d_2)$.

$t_4 = 1 - \exp(-\mu_0 \overline{L}_0)$ is the probability for a gamma quantum which has reached the detector to be registered. $\overline{L_0}$ is the averaged geometry path of the gamma-quantum through the detector

$$\overline{L}_0(k_0) = 2k_0 R_0 B(k_0)/\arcsin k_0$$

where $B(k_0)$ is complete elliptic integral of second kind [5].

If the source is continuous the formula (2) can be applied to an elementary volume dv and then T_s is given by

$$T_s = \int_v t_1 t_2 t_3 t_4 \, dv = 2\pi R_0^3 \int_0^{R_0/R_1} \frac{1}{K_0^4}\left(1 - \sqrt{1 - k_0^2}\right) e^{-\mu_1 \overline{L}_1} e^{-\mu_2 d_2} \times \left(1 - e^{-\mu_0 \overline{L_0}}\right) dk_0 \tag{4}$$

The integration being carried out over an infinite volume of the source.

To calculate the photo-part we assume that the electrons, born in the detector as a result of photo- or Compton-effect or creation of an electron-positron pair give all their energy in the sensitive volume. For detector volumes from tens to hundreds of cubic centimeters and for energies up to 3 MeV this assumption is well justified.

The photo-part P_f is calculated as the limit of two converging sequences of possibilities for successive interactions of the same gamma quantum:

$$P_f^{i-} = P_f^{(i-1)-} + (fg_i + p_i) \prod_{n=1}^{i-1} c_n$$

$$P_f^{i+} = P_f^{i-} + \prod_{n=1}^{i} c_n \tag{5}$$

where p_f^{i+} and P_f^{i-} are the overestimated and the underestimated values for the photo-part after i-fold interaction of the gamma quantum, p_i, c_i and g_i are the corresponding probabilities for interaction through photoeffect, Compton-effect and generating of $e^- - e^+$ pair for a photon with energy which it posesses after $(i - 1)$-fold Compton scattering, f is the probability for registration of the annihilation quantum. In order to determine the overestimated value of the photo-part we assume that all the photons after their $(i - 1)$-fold *Compton* scattering in the detector are absorbed through photo-effect. To obtain the underestimated value we assume that all the photons after $(i - 1)$-fold scattering leave the detector without other interactions.

A FORTRAN IV program was made to perform the calculations following the proposed model. An example for calculation of the photosensitivity of a Ge(Li) detector with a volume 40 cm³ for a source at a distance of 27 mm from the center of the sphere is shown with a solid line on Fig. 1. For comparison with dashed line is shown the relative photosensitivity of the same detector, measured by *Vylov* et al. [6]. The latter is normalized at the end of the energy range (3 MeV).

It is seen, that the spherical model describes relatively well the energy dependence of the detectors of gamma rays for energies over 200 keV. However, additional assumptions are necessary to explain the complex behaviour of the experimentally measured sensitivity at lower energies.

3. Mathematical stabilization of the energy calibration

The direct gamma spectrometry is connected with long measurements because of the low specific activity being registered. This imposes especially high requirements to the stability of the gamma spectrometric equipment. Besides the electronic methods for stabilizing of the energy calibration in the case of direct gamma spectrometry numerical methods have to be used [7]. They consist of division of the measuring time for one spectrum into several intervals during which it is supposed the calibration of the

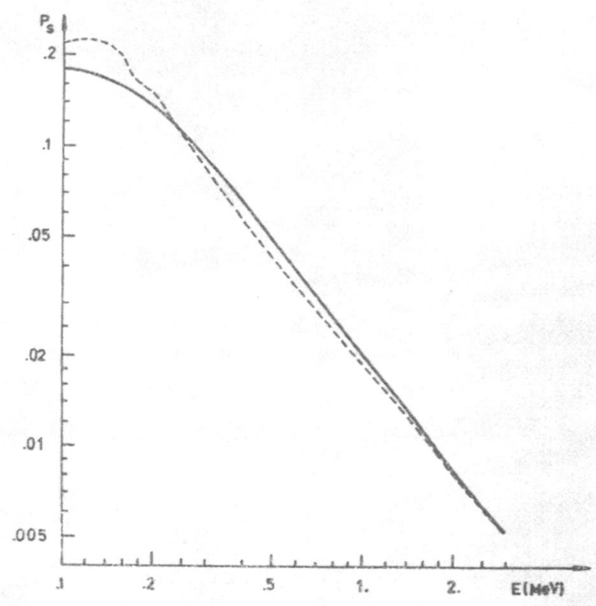

Fig. 1. A comparison between the theoretical photosensitivity of a 40 cm³ Ge(Li) detector accirding to the spherical model (thick line) and the experimental one (dashed line)

apparatus to be approximately constant. The positions of the best defined lines are determinated and used for calculation of the instrumental calibration curve for the time interval of the measurement. The scaling of the spectrum from the actual calibration curve to the standard one is preferable to be done by the filtration method. The theory of the filtration is well developed in the radio engeneering, the meteorology, the theory of the reconstitution and identification of images, where it gives very good results. However this theory is actually not used in the analysis of gamma-spectra.

The filtration is carried out by the use of the convolution integral of the following type:

$$\tilde{u}(x) = \int_{-\infty}^{\infty} h(x - x_1)\, u(x_1)\, dx_1 \tag{6}$$

where $u(x_1)$ is the experimental spectrum, $h(x - x_1)$ is the transformation function of the filter (or nucleus of the filter), $\tilde{u}(x)$ — the smoothed (filtered) spectrum. In the filtration theory it is proved that an optimal filtar for best signal to noise ratio is a filter whose form repeats the form of the useful signal. In the case of gamma lines which have a *Gaussian* form when the apparatus works correctly we use a *Gaussian* filter. It is easy to prove that the filtered line will keep its *Gaussian* form with a dispersion $\varkappa_f^2 = \varkappa_0^2 + \varkappa^2$, where $\varkappa_0^2$ is the dispersion of the experimental line and $\varkappa^2$ — of the filter's nucleus. The meansquare error σ_f of the individual points is determined from

$$\sigma_f = \frac{\sigma_0}{\sqrt{2\pi\varkappa}} \operatorname{erf}(Z) \tag{7}$$

where σ_0 is the mean-square error of the values of the non-filtered line, erf(Z) — the probability integral ($z = 2\pi\varkappa^2$).

The change of the energy scale is made by the substitution of the variables $x^* = b + ax$ in the convolution integral (6) (in the case of a linear scale of calibration), the coefficients a and b being determined from the coefficients of the real and standard calibration curves.

The numerical calculations carried out on modeled spectra confirm the theoretical predictions. In Fig. 2 an example is shown of two Gaussian curves on a constant piedestal and with a normal mean-square error. Beneath them the result of the filtration is shown using different dispersions of the filter's nucleus.

4. An estimate of the sensitivity of the method

The specific activity a for a given gamma-radioactive nuclide can be determined by the formula

$$a = 10^6\, I_{Ph}/p\alpha \tag{8}$$

where I_{Ph} is the count rate in the photopeak in counts/s, p — the photosensitivity of the detector, α — the probability for gamma quantum emission per decay.

We obtain the following expression for the mean-square error $\delta a/a$ of the specific activity a:

$$\delta a/a = 10^8 \sqrt{(I_{Ph} + 2I_{bg})/t}/ap\alpha \tag{9}$$

where I_{bg} is the count rate for the background, t — the time.

The sensitivity of the method can be defined as the minimum specific activity a_{min} which gives a photo-peak count rate equal to 3 times the mean-square error of the count rate:

$$a_{min} = 3.10^6 \sqrt{I_{bg}/t}/p\alpha \tag{10}$$

where the notation is the same, used above.

The numerical estimate of the sensitivity of the method for direct spectrometry of natural waters is made on the assumption of using a scintillation detector with dimensions ∅ 76 × 50 mm and time of measurement 72 h. It shows that about 100 gamma-radioactive nuclides can be registered with a sensitivity better than 40 mBq/l (1 pCi/l). With the use of a Ge(Li) detector with relative efficiency of 30% the theoreticaly estimated sensitivity will be improved 4 times with a simultaneous significant facilitation of the nuclide identification.

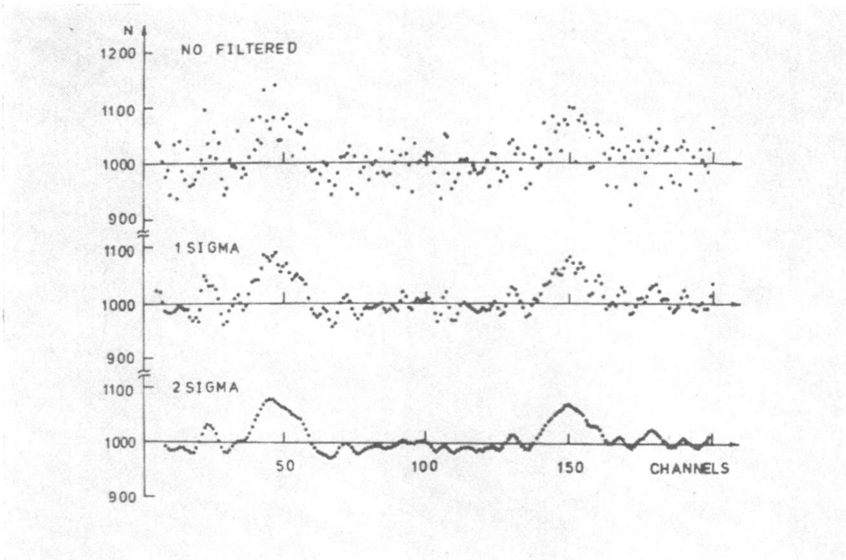

Fig. 2. The effect of the digital filtration on two Gaussians with constant background and a normal statistical error

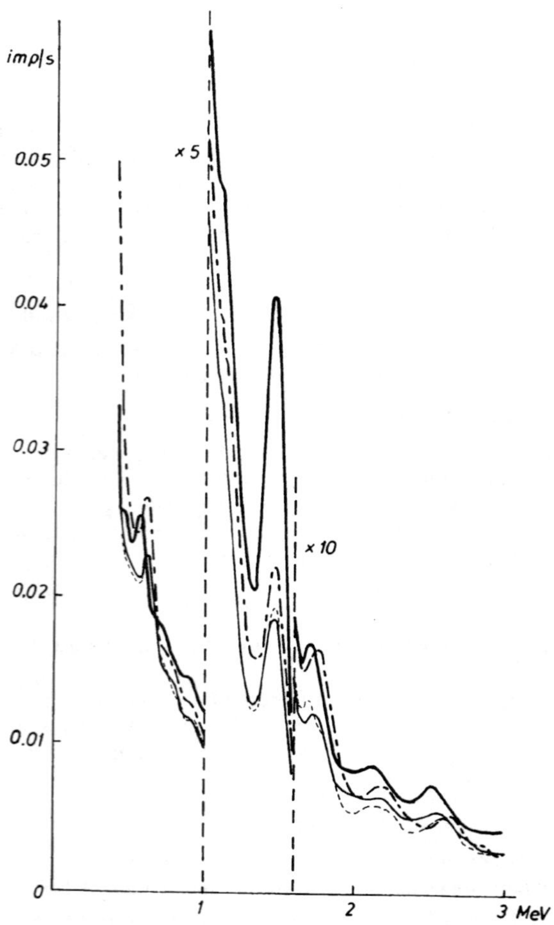

Fig. 3. spectrum of the gamma-field in different basins, registrated at depth 10 m

The digits correspond to the basins: 1 – Pirin (———); 2 – dam lake (- - - - - - -); 3 – Rila (– – – -); 4 – Black sea (▬▬▬)

In Fig. 3 are shown the gamma spectra, registered in different water basins in Bulgaria: 1. high mountain lake in Pirin, 2. dam at a medium altitude, 3. high mountain lake in Rila, 4. the Black sea. These spectra are not corrected for the detector system's own background. One can suppose that the component with energy of 1461 keV (^{40}K) in fresh waters comes mainly from this background. Probably some part of the other components comes also from radioactive nuclieds included in the materials of the water-tight container.

The presence of well marked maxima (different in the different basins) shows that there is a noticable propability, despite the strong incoherent interaction in the water, that a given part from the gamma quanta reach the detector system without interaction, which permit nuclides identification.

5. Conclusion

The method of direct gamma spectrometry of large volumes of natural waters by means of a submerged detector has potential advantages for realization of permanent automatic control of the radiation purity.

It is shown, that the principal problems connected with the calibration and the determination of the absolute concentration of the different gamma-nuclides can be solved comparably easy.

A practically sufficient sensitivity of the method is shown, as differences in the gamma field in natural water basins in Bulgaria were registered.

All this permits to us to consider that it is advisable intensity to develop further the method, aiming at its wide application.

Received January 5, 1983
Accepted in revised form April 5, 1983

References

[1] *E. Eder*, Kerntechnik **19** (1977) 21
[2] *Н. А. Вартанов, П. С. Самойлов*, Прикладная сцинтилляционная гамма-спектрометрия. Москва: Атомиздат 1975
[3] *C. Birattari, A. Salomone*, Nucl. Instr. Meth. **174** (1980) 39
[4] *R. Rieppo*, Nucl. Instr. Meth. **179** (1981) 155
[5] *Е. Янке, Ф. Эмде, Ф. Леш*, Специальные функции. Москва: Наука 1968
[6] *Ц. Вылов, Б. Н. Осипенко, В. М. Чумин*, в: Физика элементарных частиц и атомного ядра, т. 9. вып. 6, Дубна 1978
[7] *Е. Л. Столярова* и др., в: Прикладная ядерная спектроскопия, вып. 10, Москва: Атомиздат 1981, с. 149

Isotopenpraxis 19 (1983) 11, S. 391 bis 395

The Use of Radioactive Isotopes for the Study of the Metabolism of Certain Biomedically Essential Elements

I. Metabolism of Iron

A. M. El-Kot

(Chemistry Department, Faculty of Science Zagazig University, Zagazig, EGYPT)[1])

The use of radioactive isotopes in the clarifaction of the metabolism of some elements play an important role in the life cycle. A systematic study is presented on the extraction of ionic iron from urine. As radioactive isotopes, ^{59}Fe was used to follow the extraction of the ionic forms in various extraction systems. For this study we used tributyl phosphate (TBP) as solvent. Such studies are useful to develop procedures for the separation and analysis of these elements such as for example in the study of the treatment of anaemia by the administration of their compounds in the human body.

Der Einsatz von radioaktiven Isotopen zur Ermittlung des Metabolismus einiger Elemente spielt im Lebenszyklus eine große Rolle. Es wird eine systematische Studie über die Extraktion ionischen Eisens aus Harn vorgestellt. ^{59}Fe wurde als radioaktives Isotop zur Verfolgung des Extraktionsprozesses der ionischen Formen in verschiedenen Extraktionssystemen verwendet. Für diese Untersuchung wurde Tributylphosphat (TBP) als Lösungsmittel eingesetzt. Derartige Untersuchungen sind nützlich für die Entwicklung von Verfahren für die Trennung und Analysierung dieser Elemente, z. B. für die Untersuchung der Behandlung von Anämie durch die Applikation ihrer Verbindungen im menschlichen Körper.

Keywords

anemias; ascorbic acid; excretion; iron 59; metabolism; urine

Introduction

The use of radioactive isotopes in the clarifaction of the metabolism of some elements playing an important role in the life cycle, the vital problem in some countries is the anaemia diseases due to iron deficiency in the diets. Iron is the most abundant heavy metal in the earth's crust and accounts for some 4·5 per cent of all the elements; nevertheless anaemia is prevalent throughout the world and is responsible for a quite incalculable amount of ill health; its treatment is theoretically simple but in practice is often deplorable [1].

The total body iron in man is about 3·5 g of which some 70 per cent is incorporated in haemoglobin. It is usually stated that in a 70 kg man with a muscle mass of 28 kg there is only 0·14 g of iron present in the single haem group of myoglobin (muscle haemoglobin) but *Boulton* [2] has reassessed this and concluded that the amount of iron is 0·4 to 0·5 g, which amounts to 12 to 14 per cent of the total body iron. Only some 0·2 per cent of the total iron is incorporated in the vital oxidative haem enzymes such as the cytochromes, catalase and peroxidase. Most of the remaining iron, amounting in man to some 1·0 to 1·5 g is stored in the forms of ferritin or haemosiderin. A small amount is present as a "labile iron pool" and some 3 mg is active transport iron which is bound to the specific iron binding protein transferrin.

Iron is supposed to be absorbed in the ferrous state (Fe^{++}), although most of dietary iron is in the ferric form (Fe^{+++}). Only a trace of iron is absorbed in the stomach. The gastric secretions dissolve iron and promote its reduction to the ferrous state. It has been, therefore considered that ascorbic acid and other reducing substances contained in the diet help the conversion of ferric to ferrous iron and hence increase its absorption [3]. Alternatively, it has been indicated that anaemia occurs in conditions associated with ascorbic acid deficiency [4]. It has also been found that ascorbic acid metabolism is closely linked with that of iron [5, 6]. Thus it is fairly generally accepted that ascorbic acid enhances the absorption of iron [6–8].

Urinary losses in health are very small and were estimated to be less than 0·1 by *Dagg* et al. [9]. In certain haemolytic anaemias there may be considerable urinary iron losses, mainly as haemosiderin though some ferritin may be present [10]. Urinary excretion of iron – binding protein in the nephrotic syndrome has been demonstrated by *Rifkind* et al. [11].

Experimental

Apparatus

Beckmann Gamma 310 radiation counter is solid-state gamma and X-ray spectrometer counter. It is designed to measure up to 300 samples in one, two or three energy channels and report results as counts per minute (cpm), or total counts, through the Model ASR 33 Teletypewriter print out. The instrument use the detector of sodium iodide crystal activated with thallium iodide, NaI(TlI).

Radioactive Tracer

The stock solution of ^{59}Fe in the form of $FeCl_3$ ($T_{1/2} = 45$ d) was brought from New England Nuclear Company (Boston England) with specific activity equal to 8·9 mCi/mg and concentration equal to 7·0 mCi/ml. Further dilutions were made with the corresponding acid solutions.

[1]) Address of Correspondence: Chemistry Department, College of Science, King Saud University, P.O. Box 2455, Riyadh, Saudi Arabia

Reagents

Tri-n-butylphosphate (TBP), a product of Fluka A.G. Buchs S.G. ($d_4^{20} = 0{\cdot}976$) was used after purification from the mono and di-esters by washing with 5% sodium carbonate solution and then with water [12].

Nonpolar solvents, reagents grade benzene, toluene were used in these studies.

Acids of analytical grade BDH or Merck were used. $FeCl_3$, $CaCl_2$ and NaCl (Merck) were used. Samples of urine were taken from normal man. Double distilled water was used in the preparation of the aqueous solutions.

Procedure

All extraction experiments were carried out by mixing equal volumes of the organic and aqueous phases to which 50 μl of the tracer solution was added. The shaking of the mixture was carried out at room temperature (21 $\pm$ 4 °C) for about 45 min. The layers were separated by centrifugation. The distribution ratio D was taken to be the ratio of the activities of the tracer in the organic to that in the aqueous phases.

Results and Discussion

The extraction of iron from hydrochloric acid solutions has been studied by various concentrations of TBP in the organic phase and in the presence of different concentrations of urine in the aqueous phase. The results are shown in (Fig. 1–4) for 5, 10, 20 and 30% TBP in toluene, respectively. It is clear that in general the presence of urine brings about the decrease of extraction, particularly at low solvent concentrations. However, this picture is sometimes disturbed, since urine contains salts which may in some cases lead to an increase of extraction due to the salting out effect. There is a general trend of extraction, it increases with the increase of acidity in the aqueous phase. This can be explained by the fact that, according to the law of mass action, the increase of the chloride concentration in the aqueous phase should enhance the formation of the extract-

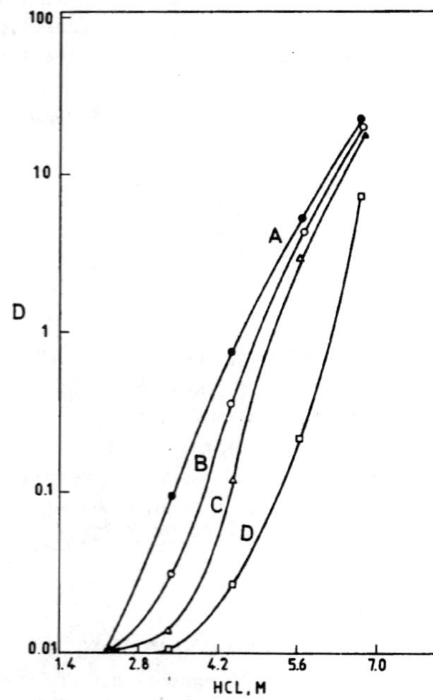

Fig. 1. Effect of HCl Concentration on the Extraction of Fe(III) by 5% TBP in presence of: 10% urine (curve A), 20% urine (curve B), 30% urine (curve C) and 40% urine (curve D)

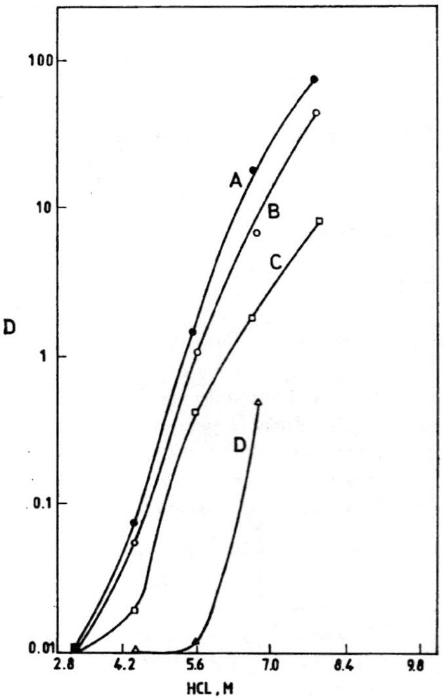

Fig. 2. Effect of HCl Concentration on the Extraction of Fe(III) by 10% TBP in presence of: 10% urine (curve A), 20% urine (curve B), 30% urine (curve C) and 40% urine (curve D)

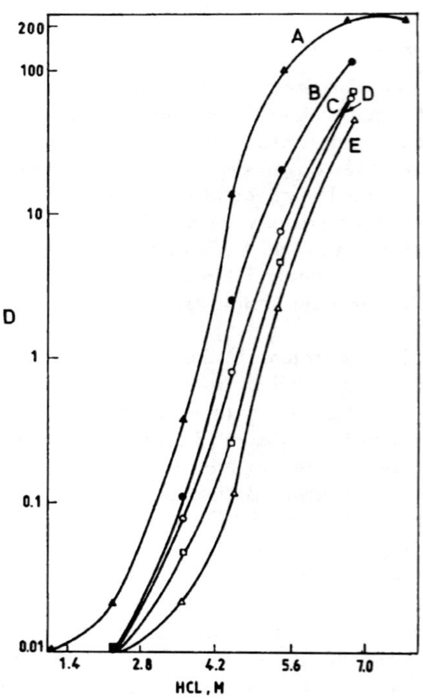

Fig. 3. Effect of HCl Concentration on the Extraction of Fe(III) by 20% TBP in presence of: no urine (curve A), 10% urine (curve B), 20% urine (curve C), 30% urine (curve D) and 40% urine (curve E)

able iron complexes. Similarly extraction increases with increase of TBP concentration in the organic phase as expected from previous studies on iron by TBP. Thus *Specker* and *Gremer* [13] described the solvent extraction of Fe(III) from HCl with 1–5% TBP in benzene. They concluded from these studies that maximum partition coefficient value is obtained from 9 M HCl. *Aven* and *Freiser* [14]

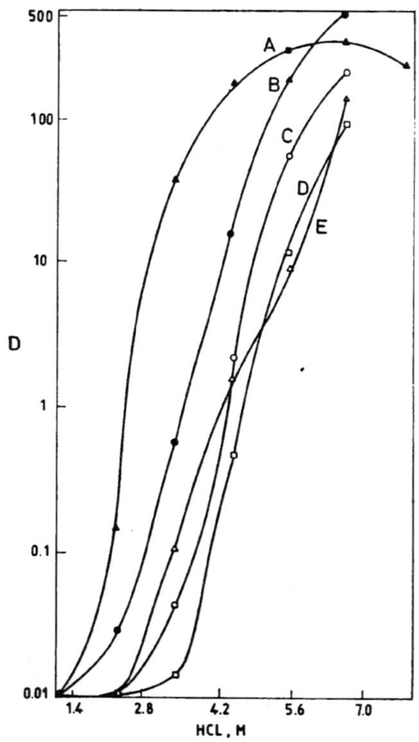

Fig. 4. Effect of HCl Concentration on the Extraction of Fe(III) by 30 % TBP in presence of: no urine (curve A), 10 % urine (curve B), 20 % urine (curve C), 30 % urine (curveD) and 40 % urine (curve E)

reported the removal of iron interference in steel analysis by extraction of ferric thiocyanate with TBP. Their work was extended by *Melnick* et al. [15], to the extraction of Fe(III) using a TBP-CCl_4 mixture. It was found by *Majumdar* and *De* [16] that almost quantitative extraction of Fe(III) is possible from about 2 M HCl using undiluted TBP (100%). *Weidmann* [17] investigated the variation of the partition of $FeCl_3$ between undiluted TBP and aqueous HCl as a function of HCl concentration. He reported that maximum extraction occurs at about 2 M HCl. *Musial* et al. [18] used 70% TBP in methanol to extract Fe(III) from HCl medium. In these conditions maximum extraction occurs at 4 M HCl.

The extraction behaviour of iron or any other metal chlorides into TBP can be formulated as:

$$M^{+m}_{aq} + mCl^-_{aq} + nTBP_{org} \rightleftharpoons MCl_m \times nTBP_{org} \quad (1)$$

where m is the charge of the metal ion and n is the number of TBP molecules solvating the cloride molecules in the organic phase. The equilibrium constant, neglecting activity coefficients, is then given by:

$$K = \frac{(MCl_m \times nTBP)_{org}}{(M^{+m})_{aq}\,(Cl^-)^m_{aq}\,(TBP)^n_{org}} \quad (2)$$

On the assumption that the activity coefficient of the organic phase is essentially unchanged, the distribution coefficient is given by the equation:

$$D = \frac{(MCl_m \times nTBP)_{org}}{(M^{+m})_{aq}} \quad (3)$$

thus equation (2) can be rewritten as:

$$K = \frac{D}{(Cl^-)^m_{aq}\,(TBP)^n_{org}} \quad (4)$$

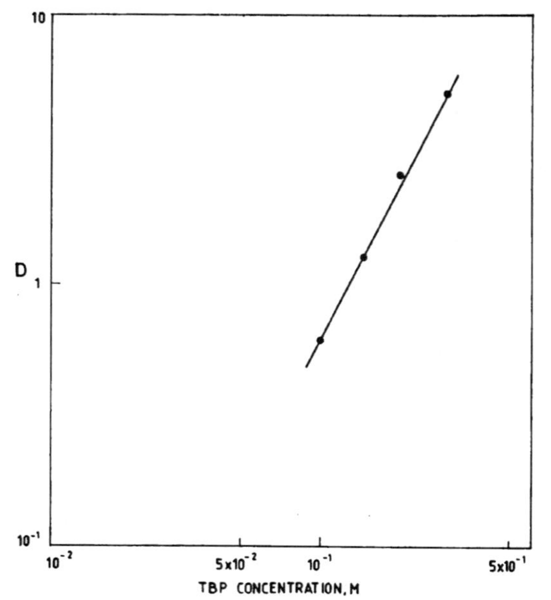

Fig. 5. Variation of D for Fe(III) with TBP Concentration at 6 M HCl

Taking the logarithm of both sides of equation (4) and at constant chloride concentration, we get:

$$\text{Log } D = n \log (TBP)_{org} + \text{constant} \quad (5)$$

If we plot log D vs. log $(TBP)_{org}$, we get a straight line of slope equal to n which is denoting the solvation number. This is illustrated by Fig. 5 which gives the plot of log D vs. log TBP concentration at 6 M HCl [19]. The best line through the points indicates a slope of 2, indicating the formation of disolvate in the organic phase. This finding confirms the data obtained by *Specker* and *Gremer* [13] who reported that the extracted species can be represented as $H[FeCl_4(TBP)_2]$ in the HCl range from 6—9 M. This result also agrees fairly well with that of *Majumder* and *De* [16] who reported that the plot of log D vs. log (TBP) at 6 M HCl gives a slope of 1·7, suggesting the extractable species to be disolvate.

Fig. 6 shows the effect of urine content on the extraction of iron by different concentrations of TBP and different

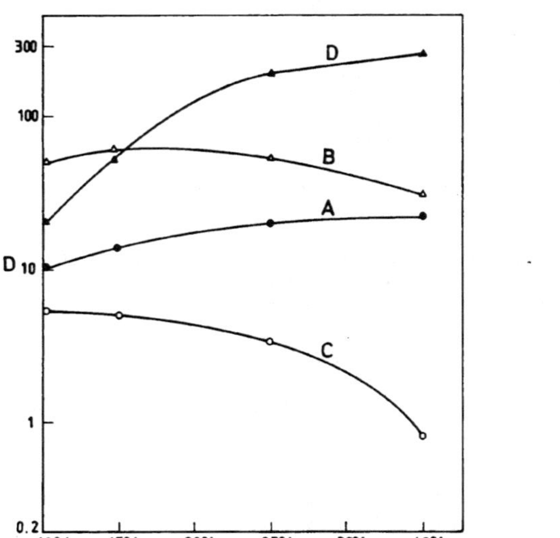

Fig. 6. Effect of Urine Content on the Extraction of Fe(III) by: 5 % TBP-6·7 M HCl (curve A), 10 % TBP-6·7 M HCl (curve B), 20 % TBP-4·4 M HCl (curve C) and 30 % TBP-5·6 M HCl (curve D)

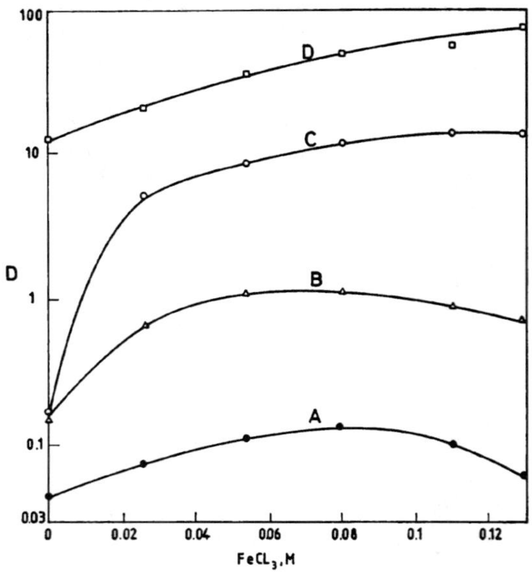

Fig. 7. Effect of $FeCl_3$ concentration on the Extraction of Fe(III) in presence of 25 % urine and 5·6 M HCl by 5 % TBP (curve A), 10 % TBP (curve B), 20 % TBP (curve C) and 30 % TBP (curve D)

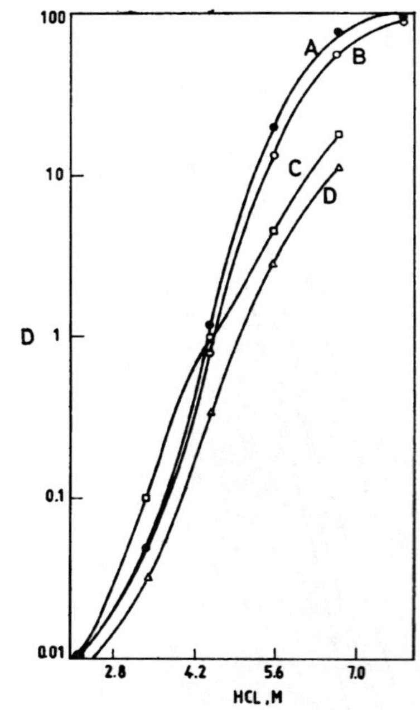

Fig. 9. Effect of Concentration of HCl on the Extraction of Fe(III) by 10 % TBP in presence of 0·033 M $FeCl_3$ and urine: 10 % urine (curve A), 20 % urine (curve B), 30 % urine (curve C) and 40 % urine (curve D)

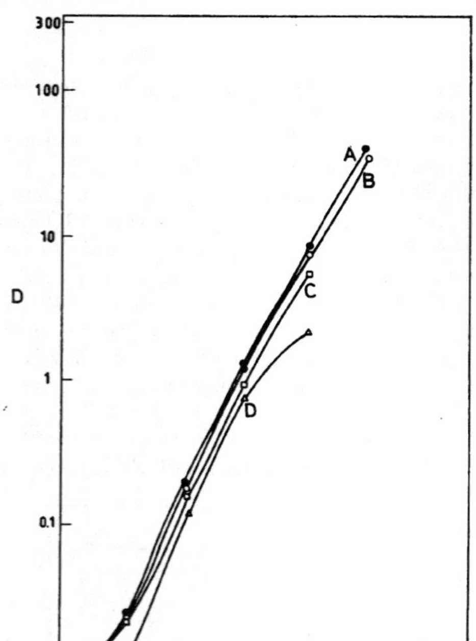

Fig. 8. Effect of HCl on the Extraction of Fe(III) by 5 % TBP in presence of 0·033 M $FeCl_3$ and urine: 10 % urine (curve A), 20 % urine (curve B), 30 % urine (curve C) and 40 % urine (curve D)

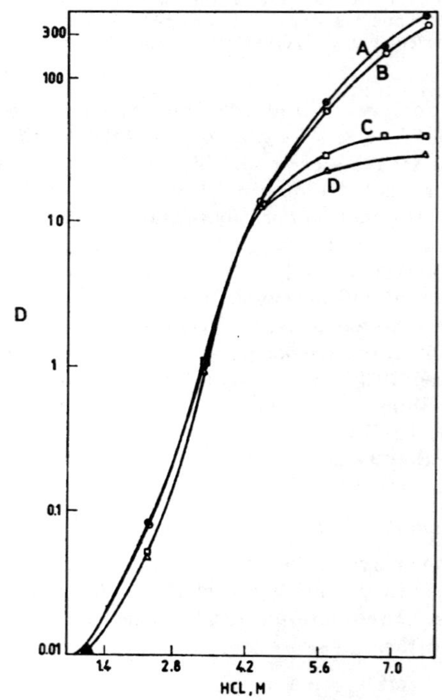

Fig. 10. Effect of HCl concentration on the Extraction of Fe(III) by 20 % TBP in presence of 0·033 M $FeCl_3$ and urine: 10 % urine (curve A), 20 % urine (curve B), 30 % urine (curve C) and 40 % urine (curve D)

concentrations of HCl. It is clear that the urine has little effect on extraction, so that extraction of iron can be carried out in presence of 40% urine in the aqueous phase. In this case, however, it is preferred that the HCl concentration is above 5·6 M.

The effect of carrier has been studied by increasing the iron concentration in the aqueous phase for extraction by various TBP concentrations from 5·6 M HCl solution (Fig. 7) It is clear that at the beginning extraction increase, with increase of $FeCl_3$ concentration due to the salting out effect of the chloride ion. However, this effect decreases at higher concentration due to the saturation of the solvent.

The extraction of iron in presence of 0·033 M iron as carrier from hydrochloric acid solutions has been also studied by various concentrations of TBP in the organic phase and in the presence of different concentrations of urine in the aqueous phase. The results are shown in Fig. 8—11 for 5, 10, 20 and 30% TBP in toluene, respectively. There is in

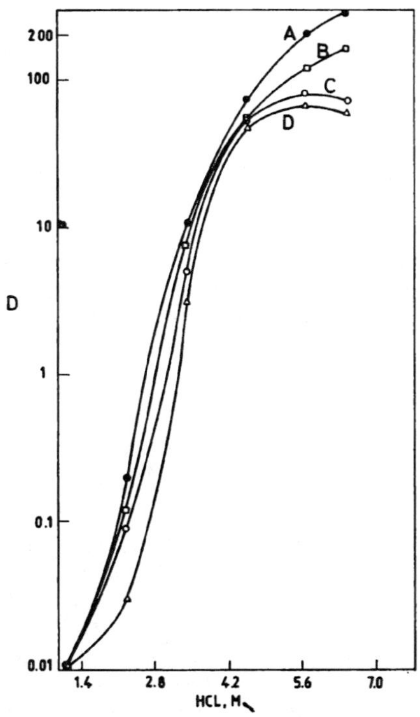

Fig. 11. Effect of HCl concentration on the Extraction of Fe(III) by 30 % TBP in presence of 0·033 M $FeCl_3$ and urine: 10 % urine (curve A), 20 % urine (curve B), 30 % urine (curve C) and 40 % urine (curve D)

general an increase of extraction due to the presence of 0·33 M $FeCl_3$ as carrier for all the different concentration of TBP.

The effect of sodium chloride concentration on the extraction of iron from 5·6 M HCl by 5, 10 and 20% TBP in toluene and presence of 25% urine, is demonstrated in (Tab. 1).

Tab. 1. Effect of NaCl on the extraction of Fe(III)

NaCl concentration	Distribution coefficient *D*		
M	5% TBP	10% TBP	20% TBP
0·0	0·034	0·18	0·87
0·08	0·044	0·23	1·50
0·16	0·050	0·34	1·50
0·24	0·070	0·14	1·51
0·36	0·080	0·21	1·63
0·40	0·070	0·24	2·30

A slight increase in the extraction of Fe(III) with the increase of NaCl concentration in the aqueous phase is observed for extraction with the three concentrations of TBP.

Also the effect of calcium chloride concentration on the extraction of iron from 5·6 M HCl by 5 and 10% TBP in toluene, and presence of 25% urine has been investigated. The results are shown in (Tab. 2). Here also only slight changes of Fe(III) extraction takes place.

Effect of Ascorbic acid on the Ectraction of Fe(III): The extraction of Fe(III) can be used as a measure of the stability of ascorbic acid, in case the latter is stable in a giyen medium it would reduce the Fe(III) to Fe(II) and thus hinders its extraction, provided that Fe(II) is not extractable under chosen experimental conditions.

Tab. 2. Effect of $CaCl_2$ on the extraction of Fe(III)

$CaCl_2$ concentration	*D*	
M	5% TBP	10% TBP
0·0	0·035	0·18
0·04	0·051	0·31
0·08	0·051	0·12
0·12	0·093	0·44
0·16	0·050	0·17
0·20	0·073	0·25

The effect of ascorbic acid concentration on the extraction of iron originally present as Fe(III) from 5·6 M HCl by 30% TBP is demonstrated from the results presented in (Tab. 3).

Tab. 3. Effect of Ascorbic Acid on the extraction of Fe(III) from 5·6 M HCl

Ascorbic acid concentration M	*D*
0·0	36·865
0·001	28·380
0·002	27·840
0·003	22·860
0·004	28·540
0·005	28·560

The above results show that ascorbic acid is unstable in 5·6 M HCl otherwise it would have reduced Fe(III) to Fe(II).

Received February 16, 1983

References

[1] *B. B. Thompson*, Disorder of the Blood. A text book of clinical Haematology Forward by C. J. C. Britton, Churchil Livingstone, Edinburgh. London and New York 1977
[2] *F. E. Boulton*, British J. of Haematology **25** (1973) 281
[3] *E. J. Underwood*, Trace Elements in Human and Animal Nutrition. Ed. 3, New York: Academic press 1971
[4] *I. J. T. Davies*, The Clinical Significance of the Essential Biological Metals, Heinemann 1972
[5] *L. E. Gunther*, Inorganic Biochemistry, volume 1/2 (1973)
[6] *E. B. Dowdle, D. Schachter, H. Schenker*, J. Physiol. **198** (1960) 609
[7] *T. H. Bothwell*, Br. J. Haematol. **14** (1968) 453
[8] *T. H. Bothwell, R. W. Charlton*, Ann. Rev. Med. **21** (1970) 145
[9] *J. H. Dagg, J. A. Smith, A. Goldberg*, Clinical Science **30** (1966) 445
[10] *D. A. Sears, P. R. Anderson, A. L. Fay, H. F. Williams, W. H. Rosby.*, Blood **28** (1966) 708
[11] *D. Rifkind, H. M. Kravetz, V. Knight*, New England J. of Medicine **265** (1961) 115
[12] *H. A. C. McKay*, Proceeding 1st cong. on Peaceful Uses of Atomic Energy, Geneva **1955**, 441
[13] *H. Specker, M. Gremer*, Z. Analyt. chem. **167** (1959) 110
[14] *M. Aven, H. Freiser*, Analyt. Chem. Acta **6** (1952) 412
[15] *L. Melnick, H. Freiser, H. F. Beeghly*, Analyt. Chem. **25** (1953) 856
[16] *S. K. Mojumdar, A. K. De*, Talanta **7** (1960) 1
[17] *G. Weidmann*, Can. J. Chem. **38** (1960) 459
[18] *A. Musial, W. Haos, G. Weidmann*, Michrochim. Acta **1962**, 883
[19] *S. A. El-Reefy*, M. Sc. Thesis, Faculty of Science, Cairo University 1972

Isotopenpraxis 19 (1983) 11, S. 396 bis 400

Buchbesprechungen

World Survey of Major Activities in Controlled Fusion Research

1982 Edition. D. Twersky (Ed.) Special Suppl. to "Nuclear Fusion", STI/PUB/23, Vienna: IAEA 1982, 420 S. (englisch) broschiert öS 700,–

Diese nun schon zum vierten Mal von der IAEA in Wien herausgegebene Weltübersicht über das Gebiet der gesteuerten Kernfusion enthält umfangreiche Informationen über die meisten Laboratorien, deren Forschungsrichtungen und das wissenschaftliche Personal. Insgesamt konnten 189 Laboratorien aus 33 Mitgliedsländern der IAEA sowie 2 internationale Organisationen (IAEA, CEC) in diesen Band einbezogen werden. Dabei waren die Herausgeber ganz auf die Bereitschaft der Mitgliedsländer und Laboratorien zur Mitarbeit und rechtzeitigen Bereitstellung der Informationen angewiesen. Das in Form dieses Bandes vorliegende Ergebnis demonstriert durch seinen Umfang und die Menge der enthaltenen Informationen deutlich das allgemeine Interesse, das einem solchen Nachschlagewerk entgegengebracht wird.

Der Band besteht aus sechs Teilen: Teil A enthält Informationen über den Aufbau und die Organisation der an der Fusionsforschung beteiligten wissenschaftlichen Einrichtungen, deren wissenschaftliches Personal und die Leitung. Hier sind auch Adressen, Telefon- und Telex-Nummern zu finden sowie Verweise auf die in anderen Teilen des Bandes ausführlicher dargestellten wissenschaftlichen Arbeitsrichtungen.

In den Teilen B bis E sind, jeweils in Spezialgebiete zusammengefaßt (B — magnetischer Plasmaeinschluß, C — reaktororientierte Technologie, D — Trägheitseinschluß, E — Theorie und numerische Studien), die wissenschaftlich-technischen Informationen in Tabellenform angegeben.

Je nach Spezialgebiet variieren die Tabellen etwas, enthalten aber in komprimierter Form Angaben über die Versuchsanlagen und Projekte, deren Status, die Namen der Hauptbearbeiter, wesentliche Parameter und gemessene Größen sowie Forschungsschwerpunkte. Im Theorieteil E sind, davon abweichend, Angaben über Modelle, deren Dimensionen und Symmetrie sowie Lösungsmethoden, Rand- und Anfangsbedingungen enthalten. Selbstverständlich können diese Angaben nur Hinweise geben, die aber beispielsweise bei der Literaturarbeit eine ausgezeichnete Orientierungshilfe sind. In diesem Sinne ist auch der Teil F von besonderem Wert. Dieser letzte Teil ist ein alphabetisches Namensregister aller in den vorhergehenden Teilen erwähnten Wissenschaftler mit entsprechenden Verweisen sowie Angabe der Länder und Laboratorien.

Dieses „Who is Who" der Kernfusionsforschung ist wegen der großen Bedeutung der internationalen Kommunikation und Kooperation auf diesem Gebiet von besonderem Wert. Das Nachschlagewerk dürfte auch der Wissenschaftsplanung und -organisation von Nutzen sein und liefert einen objektiven Überblick über die weltweiten Anstrengungen zur Erschließung der gesteuerten Kernfusion als alternative, praktisch unerschöpfliche Energiequelle.

H. Wolff

Einführung in die Sicherheitstheorie

A. Meyna. München — Wien: Carl Hanser Verlag 1982; ISBN 3-446-13581-2; 237 S., 64 Abb., 28 Tab.; kart. DM 48,–

Die Wissenschaft von der Sicherheit eines Systems ist noch sehr jung und kann gegenwärtig nicht als eine geschlossene Theorie angesehen werden. Sie ruht auf der Zuverlässigkeits- und Erneuerungstheorie und hat das Ziel, für gegebene (möglichst noch in der Konstruktionsphase befindliche) Strukturen komplexer Systeme (wie z.B. Kernkraftwerke, Chemieanlagen) quantitativ vergleichbare Kennwerte der Sicherheit bzw. des Risikos für das Betreiben der Anlagen zu definieren und zu berechnen. Sicherheit ist hier als die Wahrscheinlichkeit des Ausbleibens einer Gefährdung in irgendeinem Sinn (z.B. Strahlungsschäden) innerhalb eines Zeithorizonts aufzufassen. Insofern sind Sicherheit und Zuverlässigkeit keine identischen Begriffe (bei Ausfall der Anlage arbeitet diese zwar unzuverlässig, von einem solchen Zustand muß aber nicht unbedingt eine Gefährdung ausgehen). Das abgeschaltete Kernkraftwerk befindet sich in einem sicheren aber unzuverlässigen Zustand.

Es ist das Verdienst des Autors, diese teils verwirrende Begriffswelt und die Methoden der Bestimmung von sicherheitstechnischen Größen in dem Buch geschlossen dargestellt zu haben.

In den ersten Abschnitten werden die relevanten Begriffe der Theorie scharf und einleuchtend formuliert sowie formelmäßige Definitionen für Sicherheitskenngrößen angegeben. Abgeleitet von der Zuverlässigkeitstheorie folgen wahrscheinlichkeitstheoretische Analysen einfacher Systemstrukturen (Parallel-, Serien- und gemischte Systeme), wobei in zwei Abschnitten sog. *common mode* Fehler (verschiedene Ausfallmöglichkeiten mit gemeinsamer Ursache) und das sehr häufige menschliche Fehlverhalten in „Mensch-Maschine-Systemen" besondere Berücksichtigung finden. Letzteres führt notwendigerweise zu schwer vergleichbaren psychologischen und physiologischen (zitierten) Meßergebnissen. Die folgenden Abschnitte sind komplexen Systemstrukturen gewidmet. Es erfolgt eine Gegenüberstellung der einfacheren induktiven und der anspruchsvolleren deduktiven Verfahren (von Fehlerbaumanalysen bis zu rechentechnisch aufwendigen Simulationsverfahren auf der Basis der *Bool*schen Algebra). Unter Voraussetzung zeitlich konstanter Übergangswahrscheinlichkeiten zwischen endlich vielen Systemzuständen wird schließlich das Konzept der *Markoff*schen Prozesse für die Sicherheitsanalyse geprüft, das durch Einführung fiktiver Systemzustände breiter anwendungsfähig wird. Solche Analysen führen (auch hier) folgerichtig zu der Anschauung gut zugänglichen Graphenmodellen. Schließlich werden noch Gesichtspunkte der Erneuerungstheorie eingearbeitet. Das Buch schließt mit einer Betrachtung über die subjektiven Wichtungen einmal berechneter Risikowerte und deren tatsächlicher Annehmbarkeit auch im gesellschaftlichen Bewußtsein.

Das Buch kann nicht verheimlichen, daß es eigentlich eine kaum umgearbeitete Habilitationsschrift beinhaltet. Insofern stellt es eine wirkliche „Einführung" nur für den dar, der Elemente der Wahrscheinlichkeitsrechnung, der *Bool*schen Algebra, *Markoff*scher Prozesse (sowie der *Laplace*-Transformationen) beherrscht. Das Buch stellt das benötigte „Werkzeug" nicht dar. Das mathematische Niveau ist in den einzelnen Abschnitten sehr unterschiedlich. Es wechselt von nicht eingeführten Größen bis zu scharfen mathematischen Definitionen. Insofern ist der Leserkreis schwer einzugrenzen. Der Leser muß gutwillig sein und allen Höhen und Tiefen folgen. Dabei hilft ihm ein ausführliches Literaturverzeichnis (aus den letzten 15 Jahren), das ihn aber auch bei relativ einfachen Zusammenhängen im Stich läßt, die unscharf „wie bekannt" vorausgesetzt werden.

R. Straubel

The Handbook of Environmental Chemistry
Vol. 2: Reactions and Processes, Part A

O. Hutzinger (ed.). Berlin — Heidelberg New York: Springer 1980. 307 S., 66 Abb., 27 Tab.; geb. DM 126,— ; US $ 74,40

Umweltprobleme und ihre Folgen werden im allgemeinen von seiten der Quellen her dargestellt, mögen sie natürlicher Art oder durch anthropogene Prozesse verursacht sein. Dieser Teilband widmet sich sozusagen den Senken, den negativen Quellen. In 12 unterschiedlichen Teilaspekten werden Prozesse betrachtet, die zum Abbau, zur Veränderung oder zur Ablagerung unter natürlichen Bedingungen führen. So werden Transportprozesse in der Atmosphäre, Evaporationsgeschwindigkeiten und Adsorptionsvorgänge in der Bodenzone betrachtet, neben den photochemischen Oxydationen, den mikrobiellen Metabolismen und dem Aufnahmevermögen durch Pflanzen und unter diesem gemeinsamen Gesichtspunkt diskutiert.

An diesem Teilband haben 15 Wissenschaftler mitgearbeitet und ein sehr aktuelles Werk geschaffen. Allein die große Zahl an erfaßten Literaturstellen läßt die Breite und Aktualität des Themas erkennen; es werden 1220 Zitate aufgeführt, wobei allein auf die Jahre 1975—1980 45,2% aller Zitate entfallen, mit einem Hauptanfall um 11,8% aus dem Jahre 1978. Auch ein gewisser historischer Aspekt wird dabei keineswegs vernachlässigt — so wird auch die Arbeit aus dem Jahre 1906 zitiert, in der *N. L. Söhngen* nachweisen konnte, daß *Bacillus methanicus* mit Methan als Energiequelle (unter Oxydation zu CO_2) leben kann. Eine sehr umfangreiche tabellarische Zusammenstellung ist den unterschiedlichen Reaktionstypen in unserer Umwelt gewidmet (*V. M. Mengie*, S. 247—289 mit allein 342 Literaturhinweisen). Auch dieser Teilband bietet dem Nutzer eine Fülle von Fakten, Hinweisen und Zusammenhängen, die nur mit Mühe aus der weitverstreuten Literatur aufgefunden werden könnten. Die Thematik läßt deutlich werden, daß unsere natürliche Umwelt eine Reihe von Mechanismen bereithält, um den unterschiedlichen Stoffen, die wir an sie entlassen, zu begegnen. So gesehen, ist dieser Teilband auch eine gute Einführung in das Verständnis dieser Reaktionen und Prozesse überhaupt; der lehrende Wissenschaftler in Sachen Umweltforschung wird die Zusammenstellung mit großem Nutzen zu Rate ziehen.

H. Hübner

The Handbook of Environmental Chemistry
Vol. 3: Anthropogenic Compounds, Part A

O. Hutzinger (ed.). Berlin — Heidelberg — New York: Springer 1980. 274 S., 61 Abb., 73 Tab.; geb. DM 98,— ; US $ 57,90

Am Rande der Alpen unweit von München traten im Ebersberger Forst um 1890 massenhaft Nonnenfalter auf, die sich derart vermehrten, daß sich der entnadelte Wald braun färbte. Unermüdlich fressend, wandelten die Schädlinge die frischen harzigen Fichtennadeln in Kot um, der nach Erlebnisberichten hörbar zu Boden fiel wie ein Nieselregen. Die Raupen fraßen 2800 ha Wald kahl und doch blieb inmitten der gespenstischen Szene eine saftgrüne Oase erhalten. Hier hatten ein halbes Dutzend Waldameisenvölker nicht eine Raupe am Leben gelassen. Heute würde man mit einem ganzen Arsenal der unterschiedlichsten Schädlingsbekämpfungsmittel diesem Treiben ein Ende setzen, aber kaum jemand würde sich der Waldameisen erinnern, die so offenkundig gewirkt haben.

Welche Eigenschaften die sauerstoffhaltigen chlorierten Aromaten haben und wie sie analytisch bestimmt werden können, ist eines der Kapitel in diesem Teilband, der den anthropogenen Verbindungen gewidmet ist. Die Schwermetalle Quecksilber und Kadmium, die fluorierten und chlorierten Kohlenwasserstoffe, polycyclische Aromaten, organische Farbstoffe und anorganische bzw. organische Pigmente sind weitere Kapitel der umweltbelastenden Stoffe. Den Abschluß bilden die radioaktiven Elemente, wobei speziell T, ^{85}Kr, ^{90}Sr, ^{131}J, ^{137}Cs, ^{226}Ra und $^{239/240}$Pu behandelt werden. In allen Darstellungen werden Erzeugung, Verwendung, Akkumulationen und Abbaureaktionen behandelt, wobei den biologischen/toxikologischen Effekten ein breiter Raum eingeräumt wird. Insgesamt 12 Autoren haben diese einzelnen Teilgebiete dargestellt und ihnen eine übersichtliche Form gegeben, die durch zahlreiche Tabellen und Abbildungen aufgelockert wird.

Die Literatur ist mit 1672 Zitaten berücksichtigt, wobei 40,4% der Zitate aus den Jahren 1975—1980 stammen, den Hauptteil bilden die Zitate aus dem Jahre 1978 mit 185 Literaturstellen entsprechend 11,1%. Die ältesten Literaturstellen weisen auf die Arbeiten *G. Agricolas* (1621) und *V. Biringuccis* (1540) hin, neben *Lomonossov, Priestly, Lavoisier* und *Howard.*

Erst in den letzten Jahren erinnert man sich zunehmend der ökologischen Selbstregulationsmechanismen, und eine Ausrottung der Waldameise hätte verheerende Folgen. Die wirkungsvollen Schädlingsbekämpfungsmittel vernichten nicht nur die eigentlichen Schädlinge, sondern zugleich die erwünschten Insekten. Der vorliegende Teilband ist zwar eine buchhalterische Erfassung vieler Fakten und Daten der durch Menschen erzeugten Verbindungen, regt aber auch zum Nachdenken über unser Tun und unseren Umgang mit der Natur an. Er sollte in keiner renommierten Bibliothek fehlen.

H. Hübner

Analyse von Zeitreihen

C. Chatfield; aus dem Engl.: H. Grimm. Leipzig: BSB B. G. Teubner Verlagsges. 1982; 239 S., 36 Abb.; kart. M 29,— (Ausland: M 42,—)

Zeitreihen werden bereits seit langer Zeit in den verschiedensten Bereichen registriert. Aber erst die Anwendung von geeigneten Verfahren zu ihrer Analyse und Interpretation gestattet eine effektive Nutzung der darin enthaltenen Information. Das vorliegende Buch trägt den Untertitel „Eine Einführung", mit der Absicht des Autors, „eine verständliche Einführung in ein ziemlich schwieriges Problem zu geben und die Lücke zwischen Theorie und Praxis zu schließen". Da bisher in deutscher Sprache wenig Literatur mit diesem Anliegen existiert, ist die rasche Übersetzung ins Deutsche sehr zu begrüßen. Über illustrative Beispiele wird der Leser in die Problematik eingeführt und zunächst mit einfachen beschreibenden Techniken vertraut gemacht (Kapitel 1 und 2). Als Grundlage für die später dargelegten komplizierteren Modelle werden im 3. Kapitel einige Wahrscheinlichkeitsmodelle für Zeitreihen vorgestellt.

Kapitel 4 ist der Analyse im Zeitbereich gewidmet. Es beinhaltet die Schätzung der Autokovarianz- und Autokorrelationsfunktion, und die Anpassung von Zeitreihen mittels AR-, MA-, ARMA- und ARIMA-Modellen wird vorgestellt. Kapitel 5 führt in Probleme der Vorhersage, speziell für ökonomische Probleme, ein. Kapitel 6 und 7 behandeln die Analyse im Frequenzbereich. Es werden die klassischen Schätzverfahren einschließlich der Verwendung von Fensterfunktionen und schneller *Fourier*transformation vorgestellt. Probleme der bivarianten Prozesse werden in Kapitel 8 behandelt. Der Zusammenhang der vorgestellten Verfahren mit den vor allem für den Techniker wichtigen linearen Systemen wird in Kapitel 9 hergestellt.

Der Autor geht stets vom Standpunkt des Anwenders aus, die vorliegenden Daten zu interpretieren. Dementsprechend sind eine Vielzahl illustrativer Beispiele enthalten. Um ein tieferes Eindringen zu ermöglichen, sind eine Reihe von Übungen einschließlich Lösungen enthalten. Besonders nützlich für den Leser, auch den mit der Problematik etwas vertrauten, sind Vergleiche und Diskussionen zu den vorgestellten Schätzverfahren.

Obwohl einige neuere Verfahren, wie die Analyse im Frequenzbereich mittels Maximum-Entropie-Methode nicht erwähnt werden, ist das vorliegende Buch eine gelungene Einführung in die Thematik der Zeitreihenanalyse und kann allen Interessierten empfohlen werden. Es ist eine gute Grundlage für die Anwendung dieser Methoden, wobei die genannten Verfahren bei einer umfangreicheren Nutzung nicht neu zu programmieren sind, sondern man sollte auf in vielen Rechenzentren vorhandene Softwarelösungen zurückgreifen.

M. Remer

Rechentechnik der sozialistischen Länder

Sammelband Folge 11. Moskau: Verlag Finansy i statistika 1982, 207 S., div. Abb. u. Tab. (in Russ.); kart. M 5,—

Die Artikelsammlungen der gemeinsamen Regierungskommission der sozialistischen Länder für die Zusammenarbeit auf dem Gebiet der Rechentechnik haben sich bereits einen festen Platz in den Fachbibliotheken der Rechenzentren und der Forschungszentren der elektronischen Datenverarbeitung erobert. Auch die vorliegende 11. Ausgabe dieser Serie enthält Artikel, welche für die konzeptionelle und praktische Arbeit der Spezialisten der Rechentechnik von Nutzen sein können. Es werden im wesentlichen Konzeptionen und Erfahrungen bei der Anwendung der Rechentechnik in Bulgarien, Ungarn, DDR, Polen, UdSSR und ČSSR dargestellt und neue technische und programmtechnische Mittel der ESER und der SM-Rechentechnik beschrieben. Die Artikel werden in sechs Sachgruppen zusammengefaßt: die internationale Zusammenarbeit sozialistischer Länder auf dem Gebiet der Rechentechnik, die technischen Mittel der Rechentechnik, die Programme, die Anwendung der Rechentechnik in verschiedenen Bereichen der Volkswirtschaft, die Organisation und Dienstleistungen der Rechenzentren sowie die Information über neue technische Mittel der ESER und der SM-Serie.

Für eine langfristige Einschätzung der Entwicklung der EDV in den sozialistischen Ländern scheinen zwei Beiträge der Sammlung von besonderem Wert zu sein: die „Tendenzen der Anwendung der Rechentechnik in der Volkswirtschaft sozialistischer Länder" von *N. I. Češenko* (UdSSR) und „Über die Ermittlung des Bedarfes der Volkswirtschaft der sozialistischen Länder an Mitteln der Rechentechnik" von *P. Konakčijev* (VR Bulgarien) und *A. W. Smagin* (UdSSR).

Vor der Volkswirtschaft der sozialistischen Länder steht in den 80er Jahren eine Reihe von Intensivierungsaufgaben, welche nur durch verstärkten breiten Einsatz der Rechentechnik bewältigt werden können. Zu den wichtigsten Aufgaben der Einführung der Rechentechnik gehören nach Meinung der Autoren der Sammlung folgende: Konzentration der Bemühungen bei der Lösung der Probleme der operativen Steuerung, Leitung und Planung der Produktionsprozesse, die Probleme der Kontrolle der Qualität der Erzeugnisse, Entwicklung der Wechselbeziehungen zwischen den automatisierten Systemen der Steuerung von technologischen Prozessen und den Systemen der technisch-ökonomischen Maßnahmen in Betrieben, Kombinaten und Bereichen, Entwicklung der Automatisierung der technologischen Prozesse in Richtung der Vertiefung der Spezialisierung und Kooperation der Produktion, Schaffung von technischen Mitteln der Rechentechnik mit besseren Leistungsparametern auf der Grundlage von Mikroprozessoren, Entwicklung von Automatisierungsmitteln zur Projektierung komplizierter Objekte und Systeme im Dialog, Entwicklung der Informationssysteme für die Forschung und Entwicklung, Mittel der Experimentautomatisierung und der Auswertung der Experimentdaten. Die Realisierung der genannten Aufgaben bedarf der Einleitung von Maßnahmen zur Schaffung integrierter Systeme von Datenbanken, zur Vervollkommnung der Rechnersysteme der kollektiven Nutzung, zur Entwicklung von Rechnernetzen verschiedener Ebene und Architektur, zur Automatisierung der Projektierungsarbeiten mit Datenbankunterstützung, zur breiten Nutzung der mathematischen Mittel der Modellierung im Dialog. Einen wichtigen Beitrag haben bei diesen Prozessen die Bereiche der System- und der angewandten Programmierung zu leisten.

Die Autoren der Sammlung beschreiben einzelne typische Beispiele der Anwendung der Rechentechnik in der Volkswirtschaft, wobei besondere Beachtung der Projektierung des Systems und der Einschätzung des Aufwandes und des Nutzens geschenkt wird. So z.B. berichtet *R. Liskowskij* (DDR) über ein automatisiertes System der materiell-technischen Versorgung eines Industriebetriebes, *D. Kristmann, M. Wolf* und *D. Sebastian* (DDR) über die Anwendung der Rechentechnik zur Planung und Leitung der industriellen Schweinemastanlagen, *E. Batta* und *J. Molnar* (VR Ungarn) über ein staatliches Automatisierungssystem zur Ermittlung der Ersparnisse der Bevölkerung und 10 weitere Anwendungsbeispiele aus verschiedenen Bereichen der Volkswirtschaft. Es werden ebenfalls folgende technische Mittel der EDV vorgestellt: EC-5002.02, EC-7049, CM-6905, CM-6311, CM-6302, EC-7033 und der Textautomat A-5310. Die Sammlung kann Spezialisten, die auf dem Gebiet der Entwicklung und Nutzung der Systeme ESER und CM arbeiten bzw. automatische und automatisierte Anwendungssysteme projektieren, als eine wertvolle Literaturquelle empfohlen werden.

L. Jankowski

Einführung in die Programmierung mit PASCAL

H.-E. Erbs u. O. Stolz. Stuttgart: B. G. Teubner 1982. ISBN 3-519-02506-X; 232 S.; kart. DM 22,80

Das Äußere und der Titel des Buches versprechen den üblichen Inhalt solcher „Lehrbücher": eine systematische Darstellung mit allen (zunächst unwichtigen) Feinheiten, die den Leser allein läßt, wenn er sich tatsächlich an einen Rechner wagt. Beim ersten Aufschlagen des Buches werden jedoch „naiv" gezeichnete Bildchen sichtbar, die auf eine andersgeartete Darstellung hinweisen. Das Buch bietet ein Rezept, etwa nach der Vorschrift eines „Lernbuches": Man nehme das Buch, Papier und Bleistift, sowie ein Bildschirmgerät, das mit einem „PASCAL-verstehenden" Rechner verbunden sein muß; dazu noch eine Betriebsanleitung für das Terminal. So ausgerüstet taste man sich anhand des Buches jede Woche durch (nicht mehr als) eine Lektion und wird am Ende des Buches die Spitze des „Rocky Pascalians" erreicht haben. — Das dürfte auch tatsächlich möglich sein, denn der Inhalt der Lektionen und insbesondere die dargestellten Beispiele sind didaktisch so gut gegliedert, daß jeweils nur die Sprachmittel dargestellt werden, die in der entsprechenden Lektion auch tatsächlich benötigt werden. Dabei wird der Leser niemals mit Einzelheiten überhäuft, ist aber in der Lage, schon nach dem Durcharbeiten der ersten Lektion sein erworbenes Wissen praktisch zu erproben.

Das Buch beginnt mit der Formulierung einfachster Programme, die ohne Dateneingabe auskommen. So wird zunächst eine Tischkarte mit dem Namen „OTTO" gedruckt. Danach folgen Vergleichsoperationen, Schleifen, Entscheidungen, Verschachtelungen, Standard- Datentypen, -Funktionen und -Texte, Laufanweisungen, Verbund-Datentypen, Prozeduren usw. bis hin zu schon anspruchsvollen Programmen der Behandlung großer Datenmengen etc. Zwischendurch bekommt der Leser Hinweise zur ökonomischen Programmierung, wie die Aufstellung von Struktogrammen (top down design), die Abschätzung von zeitlichen Schleifenkosten sowie die Suche nach vorteilhaften Schleifeninvarianten.

In jeder Lektion sind (meist humorige) Beispiele vollständiger Programme dargestellt, an denen die verbalen Erläuterungen des Textes sofort veranschaulicht werden. Jede Lektion schließt mit auf den gerade vermittelten Stoff direkt bezogenen Übungsaufgaben, die den Leser offensichtlich nicht nur befähigen sollen, in PASCAL zu schreiben und zu lesen, sondern auch zu denken, also aktiv zu werden, ohne den rasch zunehmenden Schwierigkeitsgrad zu bemerken; auch verhältnismäßig komplizierte Sachverhalte sind in dem Buch von (auch didaktisch erfahrenen) Praktikern allgemeinverständlich dargestellt.

Das soll natürlich nicht darüber hinwegtäuschen, daß das eigentliche Lernen auch nach diesem Buch harte Arbeit ist. So erfordert nicht nur das Lesen der Programmbeispiele große physische Aufmerksamkeit, da die verschiedenen Klammerausdrucke sich (leider) im verwendeten Druck nur geringfügig unterscheiden.

Das Buch ist aber — wie der Titel vermuten läßt — nicht nur für den Programmier-Anfänger geschrieben. Auch der in älteren Programmiersprachen denkende, versierte Programmierer dürfte beim Studium des Buches in amüsanter Weise seine Gedankenwelt erweitern können.

Auf Literatur ist nur — wo unbedingt nötig — in Fußnoten hingewiesen. Auch weiterführende Literatur wird nicht erwähnt. Dagegen besitzt das Buch ein ausgezeichnetes Sachwortverzeichnis, welches das Buch auch zu einem Nachschlagewerk erhebt; für denjenigen, der es durchgearbeitet hat.

R. Straubel

Encyclopedia of Plant Physiology, New Series Vol. 12B Physiological Plant Ecology II Water Relations and Carbon Assimilation

O. L. Lange, P. S. Nobel, C. B. Osmond, H. Ziegler (editors) Berlin — Heidelberg — New York: Springer 1982. 747 S., 153 Abb., 40 Tab.; geb. DM 288,— ; US $ 120,10

In der Botanik sind Fragestellungen und Untersuchungsmethoden zu den vielseitigen Wechselwirkungen zwischen den Pflanzen und ihrer Umwelt viel länger bekannt als in der Zoologie. Standortanalysen, Klimafaktoren und experimentelle Methoden der Pflanzenphysiologie unter Einbeziehung physikalischer und chemischer Methoden sind gewohnte Arbeitsweisen, die zu quantitativen Aussagen führen und mathematisch-formulierbare Zusammenhänge aufdecken. Wenn man — vereinfachend gesagt — so will, ist das Leben der Pflanzen an das Wasser gebunden, und ihre Reaktion darauf ist ihr Wachstum, ihr Gedeihen, die Photosynthese. Der vorliegende Teilband 12 B der Pflanzenphysiologie ist nur diesem Wechselverhältnis gewidmet: Wasser als Stressor auf die Kohlenstoffassimilation, Trokkenheit und Überflutung als begrenzende Faktoren für das Leben der Pflanzen und ihr Anpassungsverhalten darauf.

An der umfangreichen und modernen Zusammenstellung haben 26 Autoren mitgearbeitet, die in 18 Artikeln das Anliegen dieses Teilbandes darzustellen versuchen. Wasseraufnahme und Wassertransport, Wasserspeicherung und Wasserverlust, Toleranzgrenzen der Austrocknung, Wirkung von Frost, Dürre und Überflutung sind in gesonderten Kapiteln — kaum mit Überschneidungen — dargestellt und immer als ökologische Faktoren abgehandelt.

Die in den vielen Reisenotizen der Naturwissenschaftler des vergangenen Jahrhunderts dokumentierten geologischen, geographischen, klimatologischen und ökologischen Beobachtungen der Pflanzenwelt ferner Länder erfahren gegenwärtig ihre physiologischen und biophysikalisch/biochemischen Erklärungen, wobei ein beachtenswerter Apparat der Experimentiertechnik angewendet wird. Ergebnisse, die mit Hilfe von radioaktiven Isotopen (^{14}C-markierten Verbindungen) und durch Interpretation der natürlichen Isotopenvariationen (^{12}C/^{13}C- und H/D-Verhältnisse) erhalten wurden, werden in mehreren Beiträgen angeführt. Die natürliche Selbstverständlichkeit, mit der die Aussagen der Isotopenmethoden zur Erklärung herangezogen werden, ist ein schöner Beweis dafür, wie gewohnt diese Arbeitstechnik in der Pflanzenphysiologie heute schon ist.

Es ist fast eine Trivialität, festzustellen, daß Wissenschaftler, die durch ihre Forschungen das Arbeitsgebiet selbst maßgeblich beeinflussen und vorantreiben, auch eine sehr gute Kenntnis der Literatur auf diesem Gebiet haben, bis in die jüngste Vergangenheit zurück. Im vorliegenden Band werden insgesamt 2713 Literaturstellen angegeben, wobei die für 1976—1982 allein 45,6% ausmachen, mit einem maximalen Anteil von 10,1% aus dem Jahre 1978 (entsprechend 273 Zitaten). Manche Autoren haben auch bereits Zitate aus dem Jahre 1982 (13 Zitate) — dem Erscheinungsjahr des Bandes — mit verarbeitet; demgegenüber wird aber der historische Aspekt keineswegs unterdrückt, das älteste Zitat stammt aus dem Jahre 1806 — *Alexander von Humboldt*: Ideen zu einer Physiognomik der Gewächse.

Ein weiterer Teilband der Physiologischen Pflanzenökologie liegt vor — eine ausgezeichnete Zusammenstellung und mit meisterhafter Disziplin der Einzelautoren verfaßte Berichterstattung unseres gegenwärtigen Wissens. Auch dieser Teilband wird für viele Lernende und Lehrende eine Fundgrube an Antworten und Problemen der jeweils aufgeworfenen Fragen sein.

H. Hübner

Food Irradiation Now

Proc. of a Symposium held in Ede, the Netherlands, 21. Oct. 1981. The Hague — Boston — London: Martinus Nijhoff/ Dr. W. Junk Publishers 1982. 157 S., 18 Abb., 18 Tab.; geb. Dfl 50.00; US $ 22.00

Holland ist in der Anwendung der Strahlenbehandlung von Lebensmitteln weltweit führend. Hervorragenden Anteil an der Erringung dieser Position hat die Firma Gammaster, die aufgrund des stark steigenden Bedarfs eine zweite, leistungsfähige Gamma-Bestrahlungsanlage errichtet. Gammaster förderte auch das Symposium und die Herausgabe des vorliegenden Buches, dessen Anliegen die Darlegung der Möglichkeiten und des Nutzens der Strahlenbehandlung von Lebensmitteln ist. In acht Übersichtsvorträgen werden alle Aspekte der Strahlenbehandlung von Lebensmitteln behandelt.

Durch Lebensmittel übertragene Infektionen, vor allem Salmonellosen, gewinnen in steigendem Maße in vielen Ländern an Bedeutung. *Mossel* und *Netten* (Utrecht) stellen diese Problematik unter besonderer Berücksichtigung der Belastung von Frischfleisch und Geflügel durch enteropathogene Mikroorganismen dar. Die sich durch eine Strahlenbehandlung von Lebensmitteln eröffnenden Möglichkeiten zur Verbesserung der Volksgesundheit werden von *Kampelmacher* (Wageningen) dargelegt.

Da die Strahlenbehandlung von Lebensmitteln unter den verschiedensten Zielstellungen erfolgen kann, z.B. Keimungshemmung, Reifeverzögerung, Inaktivierung von Schadinsekten und pathogenen Organismen, Pasteurisierung oder Sterilisierung, und da sowohl Elektronenbeschleuniger als auch Gamma-Strahlung zur Anwendung kommen können, sind die jeweiligen technologischen Lösungen besonders unter ökonomischen Aspekten von Interesse. *Langerak* (Wageningen) und *Leemhorst* (Ede) berichten über die technologischen Aspekte bzw. über Erfahrungen mit bereits arbeitenden Anlagen.

Für die breite Einführung der Verfahren der Strahlenbehandlung von Lebensmitteln sind sowohl nationale wie internationale Regelungen erforderlich. Einen Überblick über entsprechende Aktivitäten gehen die Beiträge von *Cornelis* (Rijswijk), *Ulmann* (The Hague) und *van Kooij* (IAEA, Wien), wobei letzterer auch auf den Stand der Zulassung dieser Verfahren in den einzelnen Ländern eingeht.

Mit der Reaktion der Verbraucher auf die Strahlenbehandlung von Lebensmitteln setzt sich *Defesche* (Amsterdam) auseinander, wobei sowohl die Vorbehalte gegenüber der Strahlenbehandlung von Lebensmitteln und deren Ursachen als auch die Vorteile des Verfahrens, z.B. im Vergleich zur chemischen Konservierung oder hinsichtlich der Qualität des Produktes, diskutiert werden.

Ergänzt werden die Vorträge durch den Abdruck der Beantwortung einer Reihe von Fragen zur Mikrobiologie, Technologie, Gesetzgebung und zu Verbraucherfragen, die mit der Strahlenbehandlung von Lebensmitteln im Zusammenhang stehen.

Außerdem ist der Bericht des gemeinsamen Expertenkomitees der FAO/IAEA/WHO zur gesundheitlichen Unbedenklichkeit strahlenbehandelter Lebensmittel (Genf, 27. Oktober bis 3. November 1980) in voller Länge abgedruckt, so daß das vorliegende Buch für alle an dieser Problematik Interessierten eine gute Einführung darstellt.

G. Hübner

ISOTOPENPRAXIS

Volume 19 | 1983 | Number 11

Contents

Articles

Inhalt

Originalarbeiten

Isotopenpraxis is indexed in Current Contents/PC & ES, Science Citation Index, and INIS Atomindex

FSC
www.fsc.org
MIX
Papier | Fördert gute Waldnutzung
FSC® C083411

Zeitfracht Medien GmbH
Ferdinand-Jühlke-Straße 7
99095 Erfurt, Deutschland
produktsicherheit@kolibri360.de